The Cellular Cosmogony

Earth is a Concave Sphere, with Proof

By

Cyrus Reed Teed

First published in 1922

Published by Left of Brain Books

Copyright © 2023 Left of Brain Books

ISBN 978-1-397-66856-1

First Edition

PUBLISHER'S PREFACE

About the Book

"Teed, born in 1839 in upstate New York, served with the Union Army, and later became a herbalist and studied alchemy. In 1869 Teed had a vision in his laboratory, in which a beautiful woman spoke to him and revealed that he was to become a messiah, and reveal the true cosmogony to the world. Teed took the name Koresh (not to be confused with David Koresh of Waco). He preached that belief in the concavity of the Earth is equivalent to godliness. He proclaimed, "All that is opposed to Koreshanity is antichrist". After touring widely preaching Koreshanity, he settled in Chicago, and started a communal society, as well as a periodical, The Flaming Sword. Koreshanity, at its height, had a few thousand followers. In the 1890s Teed founded the town of Estero, Florida, near Ft. Meyers, and declared it the coming capital of the world. His followers prepared for eight million believers to show up. Only two hundred did.

Teed wrote a book called The Immortal Manhood, in which he taught that after his death he would rise to heaven and take all of his followers with him. Alas, when he died in 1908, nothing of the sort happened. After weeks, the county health officer finally had to step in to force the colony to bury the rotting body of their leader. Teed was buried in a tomb which was later washed out to sea in a hurricane in 1921. The Flaming Sword continued to be published until 1949, although it never mentioned that Teed had died. As a matter of fact, nowhere in this book is it mentioned that the author had been dead for fourteen years when it was published.

The Cellular Cosmogony was Teed's magnum opus. Teed propounded that the surface of the earth is concave, not convex, and that the entire universe is contained within the 25,000 mile circumference of the inside-out earth. The Sun is in the exact center of the 'cosmic egg,' 4,000 miles away, and is actually a helix. However we never see this directly, only some kind of reflection of it. The Sun is dark on one side, which produces day and night. The moon is a reflection of the Earth, and Teed believed he could see outlines of the Earth's continents and seas on it! Other astronomical phenomena are essentially optical illusions. Besides geology, he also denounces the scientific method, the Copernican theory, the atomic theory, modern chemistry, conventional surveying techniques, and last but not least, optics. Truly, 'everything you know is wrong.'

Teed and his followers devoted much time and energy to practical experiments to prove the concavity of the earth. Whether there was some deception involved, or self-deception, it is difficult to tell at this late date. Their surveying methodology and the device they used to take the measurements with (the 'rectilineator') have both been called into question.

The obvious problems with his cosmology are either not covered at all or brushed aside with a flurry of invented (and often semantically null) polysyllabic words. Why does the sun rise and set each day? What is the horizon and why is it about five miles away at sea level? How come we can't see locations hundreds of miles away just by looking up a bit?"

(Quote from sacred-texts.com)

About the Author

Cyrus Reed Teed (1839 - 1908)

"Cyrus Reed Teed (October 18, 1839 - December 22, 1908), was an eclectic physician who was the creator of a unique Hollow Earth theory and founder of the Koreshan Unity.

As a young physician, Teed was always interested in unconventional experiments, often involving dangerously high levels of electricity. One day, during an experiment he was badly shocked, and passed out. During his period of unconsciousness, Teed believed he was visited by a divine spirit who told him that he was the messiah. Inspired, once he awoke Teed vowed to apply his scientific knowledge to "redeem humanity." He promptly changed his first name to "Koresh," the Hebrew word for Cyrus.

Some believe that the heavy dose of electricity Teed absorbed may have damaged his brain. Indeed, following his "divine visit" Teed's scientific theories began to get increasingly odd. He denounced the idea that the Earth revolved around the sun and instead pioneered his own theory of the Universe, known as the Cellular Cosmogony. According to this theory, human beings live on the inside of the planet, not the outside. Gravity thus does not exist, and humans are held in place due to centrifugal force. The sun is a giant battery-operated contraption, and the stars mere refractions of its light.

Though denounced by many, Teed's ideas, called Koreshanity caught on with others. Koreshanity preached cellular cosmogony, alchemy, reincarnation, immortality, celibacy, communism, and a few other radical ideas. Teed started preaching Koreshanity in the 1870s in New York, forming the Koreshan Unity, later moving to Chicago. Teed's followers formed a commune in Chicago in 1888. Some followers also formed a short-lived community in San Francisco (1891-2). Small church groups existed in other towns.

Eventually, Teed took his followers to a small Florida town named Estero, to form his "New Jerusalem" in 1894. The 'golden age' for this community was 1903-1908, when they had over 250 residents and incorporated Estero. They built extensively, establishing a bakery, printing house (publishing their newspaper and other publications), their "World College of Life", a general store, concrete works, power plant (that supplied power to the surrounding area years before it was available elsewhere in the region) and more. The colony was extensively landscaped with exotic tropical plantings. They tried to run several candidates for county government against the local Democratic Party but were never successful. After Teed died in 1908 (after being beaten by the town marshall), the group went into decline. Because one of the beliefs of the Koreshans was in reincarnation, they thought their leader would come back to life, and instead of burying him, propped him up in a bath tub. After several days, local health officials stepped in and forced the burial of Teed.

The last members of the community deeded the colony to the State of Florida. It is now the Koreshan State Historic Site."

(Quote from wikipedia.org)

CONTENTS

INTRODUCTION

THE AUTHOR of the Koreshan System of Universology (upon the basis of the law of comparative analogy) announced, in 1870, the discovery of the cosmogonic form, which he then declared to be cellular; the surface of the earth being concave, with a curvature of about eight inches to the mile. This rate of curvation would give a diameter of eight thousand, and a circumference of twenty-five thousand miles.

Applying the common laws and principles of optics, with perspective foreshortening, all the phenomena of appearances in optical illusion were scientifically accounted for, and the earth optically demonstrated to be concave, although visual appearances seemed to indicate the contrary.

Upon the assumption that the surface of the earth is convex, there has been predicated that prodigious fallacy, the Copernican system, which, according to the admission of its most enthusiastic advocates and adherents, does not contain a single positive proof of scientific accuracy.

If we accept the logical deduction of the fallacious Copernican system of astronomy, we conclude the universe to be illimitable and incomprehensible, and its cause equally so; therefore, not only would the universe be forever beyond the reach of the intellectual perspective of human aspiration and effort, but God himself would be beyond the pale of our conception, and therefore beyond our adoration.

The Koreshan Cosmogony reduces the universe to Proportionate limits, and its cause within the comprehension of the human mind. It demonstrates the possibility of the attainment of man to his supreme, inheritance, the ultimate dominion of the universe, thus restoring him to the acme of exaltation,--the throne of the Eternal, whence he had his origin.

The alchemico-organic cosmos (the physical universe) is the ultimate and therefore the most outward expression of creative power. It is the language of causation manifest in the form of symbolism. Given a knowledge of this form with its function, the cause is necessarily disclosed.

Deity, if this be the term employed to designate the Supreme Source of being and activity, cannot be comprehended until the structure and function of the universe are absolutely known; hence mankind is ignorant of God until his handiwork is accurately deciphered. Yet to know God, who, though unknown by the world, is not unknowable, is the supreme demand of all intellectual research and development.

Embraced in the system of which the external cosmic form is the mere outward cell is its correspondent, the macrocosmic or Grand Man. Outwardly, this is the visible humanity in the process of development toward a perfected state, not yet approached only as it was attained to in the Lord Christ, its germinal beginning, the firstfruits of the perfected genus.

The alchemico-organic cell is definitely structured. It is the egg from which the human macrocosm attains its incubation; hence, when humanity reaches its organic shape it must assume the correspondential organism. Therefore, if we know universal form as it obtains in the alchemico-organic cosmos, we can render the language of this form into that of the legitimate structure of society; for one is the pattern of the other. Its functions and activities can also be readily translated into the language of societal functions and activities.

Genuine societal fellowship will finally become a structured and composite unity, evolved through the application of intellectual potency and direction. It will become the anthropomorphic counterpart of the alchemico-organic (physical) world. A knowledge of the structure and function of the alchemico-organic cosmos constitutes the basis of a structured theology, which is essential to the organic unity and perpetuity of the human race.

It does not follow, because a fallacious theological system has maintained the world in a state of offensive and defensive belligerency, anarchy, and chaos, that a true knowledge of God and his relation to man and man's relations to him is not essential to genuine human perfection and happiness. In the emphasis of the contrast, regarding the end to be attained, between the genuine (the Koreshan) and the fallacious (the Copernican) system of cosmogony, we insist that all the labor of investigation, the time, and the wealth expended in the Copernican fallacy have had no specific purpose.

Why do we care to know whether the earth moves or is stationary? If the universe be illimitable, it is equally incomprehensible. Why, then, should humanity waste its potencies in the investigation of that which it has already pronounced incomprehensible, unknown, unthinkable, because illimitable? We sought to know the exact form of the universe because we knew it to be the language of cause; and knowing the effect, we assured ourselves of the cause, that through conscious knowledge we might enter into and become its power.

The interest of the Koreshan mind in the acquisition of universal knowledge resides in the fact that through it we become intellectually instrumental in the organization of the social fabric, and thus acquire social and individual perfection, thence universal and individual immortality. We learn to know of the form and function of the alchemico-organic cosmos, that we may insure the construction of the organo-vital cosmic organism.

The perfection of the individual structure depends upon the perfection of the Grand Man. The perfection of the human macrocosm (the Grand Man) depends upon the application of a few fundamental principles, revolutionary and sweeping in their influence. The principles of equilibrium are as essential to the institution and perpetuity of human happiness as they are to the eternal stability of the cosmic structure, the basis of which knowledge is found in the shape of the surface of the earth as geometrically confirmed in the application of the Rectilineator.

As the astral nucleus of the alchemico-organic cosmos is so related in form and function to the circumference containing it, as to receive the convergence of all "energies" [1] [substances], and radiate them equitably to all parts of the cosmic structure; and as the heart of the vidual body, the seat

---

[1] Energy is not a substance, but simply the work of the two qualities which could not engage in work were there not two constant states of the two. The term energy applies equally to matter and spirit. Matter and spirit, or matter and its essence, could not be active but through their relationship; and both are active by virtue of their conjoined effort as counter-parts in the processes of the perpetuity of creation. The term energy means in work; when matter is active it is in work, and the phenomenon of motion is the energy of matter. When its co-ordinate spirit or essence is in motion it is in work; therefore, we have the in-work or the energy of spirit. It requires these three to constitute a constantly active primate cycle of being.--Koresh in Flaming Sword, Vol. XIX, No. 16.

of the commerce of the body, is the center of collection and distribution, hence society must be so organically structured as to be able to collect and distribute the products of Nature, coupled with industry and sustained by the application of economic law. The attainment of a knowledge of this law is founded upon a knowledge of the alchemico-organic cosmos.

It is so vastly important, in view of all these facts, for the world to possess a knowledge of the contour of the earth's surface, that we devoted the work of many months to mechanical application, for the purpose of giving to the world some simple mechanical proofs of what we have known and taught for [many] years.

Man's knowledge of man is his knowledge of God; not man as he is in his segregate state, but as he will be when the two forms (male and female) unite in the integralism of his biunity, of which the Son of God was the archetype. The Lord was Jehovah manifest in his human perfection.

To know the Lord Christ absolutely is to be in the consciousness of Deity; and to become like him is to sit upon the throne of his glory. This knowledge is so related to the structured alchemico-organic macrocosm, that to know of the earth's concavity and its relation to universal form is to know God; while to believe in the earth's convexity is to deny him and all his works. All that is opposed to Koreshanity is antichrist.

THE CELLULAR COSMOGONY

Koreshan Universology Is Predicated Upon an Absolutely Demonstrated Premise; the World Face to Face with a Radical Astronomical Revolution

KORESHANITY is universal science applied to all the concerns of practical life, involving the science of immortal life in the body. It includes the science of religion, founded upon an accurate knowledge of the structure and function of the cosmic organism. It embraces every department and phase of form and function in the universe, and is therefore Universology. It is predicated upon an absolutely demonstrated premise, a geometric figure which embraces three simple elements--the arc, chord, and radius, practically applied to earth measurement by a process which determines the contour of the surface of the earth in which we dwell, and the direction of its curvature. This is not theoretical, but applied geometry. This contour is found to be an upward instead of a downward curve, and thus it is diametrically opposite to the assumed convexity of the earth's surface. The world is therefore face to face with a radical astronomical revolution.

The earth is a concave sphere, the ratio of curvation being eight inches to the mile, thus giving a diameter of eight thousand, and a corresponding circumference of about twenty-five thousand miles. This fact is physically and mechanically demonstrated by placing a perpendicular post at any point on the surface of the earth, (though it were better to place it by the side of a surface of water,) and extending a straight line at right angles from this perpendicular. The line thus extended will strike the surface at any distance proportionate to the height of the vertical post.

"Hypothesis, or guesswork, indeed, lies at the foundation of all scientific knowledge," says the Standard Dictionary, quoting from Fiske's "Unseen World,". The term science is derived from scire, to know; hence, science is the Latin term for knowledge. Science means knowledge, nothing more, nothing less. That which is founded upon hypothesis (assumption) is not science, nor should it be dignified by that title. The Copernican system of astronomy has its foundation in assumption,--this is conceded by all so called scientific astronomers. The Copernican system has never been demonstrated, therefore it is not scientific.

What does Koreshanity offer as a substitute for the gigantic fallacy and farce of the benighted Copernicus? First, it offers the fact that in experiments carefully made by the Koreshan Geodetic Staff at Naples, on the Gulf coast of Florida, the contour of the earth was proven to be diametrically the reverse of what is taught as true in the pseudo-science of modern times.

The surface of the earth is not convex. It appears to be so because of optical illusion. The only geodetic survey ever made for the purpose of determining whether the surface on which we dwell is convex or concave, was made by the Geodetic Staff of the Koreshan Unity in the year 1897. In this survey was corroborated conclusively the testimony given in 1870, that the earth is a hollow shell about eight thousand miles in diameter, and about twenty-five thousand miles in circumference.

The Form of the Universe, the Great Alchemico Organic World

The alchemico-organic (physical) world or universe is a shell composed of seven metallic, five mineral, and five geologic strata, with an inner habitable surface of land and water. This inner surface, as the reader already understands, is concave. The seven metallic layers or laminæ are the seven noble metals,--gold constituting the outermost rind of the shell. This shell or crust is a number of miles in thickness.

Within this shell are three principal atmospheres, the first or outermost (the one in which we exist) being composed chiefly of oxygen and nitrogen; the one immediately above that is pure hydrogen, and the one above the hydrogen atmosphere we have denominated aboron. Within this is the solar electromagnetic atmosphere, the nucleus of which is the stellar center. In and occupying these atmospheres are the sun and stars, also the reflections called the planets and the moon.

The planets are mercurial disci moving by electromagnetic impulse between the metallic laminæ or planes of the concave shell. They are seen through penetrable rays, ultra electro-magnetic, reflected or bent back in their impingement on spheres of substance regularly graduated as the stories in the heavens.

In the foregoing is presented a descriptive outline merely, of the alchemi-co-organo-cosmic form. It is not assumptive. Neither is it intended, in this synopsis, to prove the Koreshan Universology; the proofs and demonstrations of the System will be found in subsequent pages of this work.

Motion and Function

We have already outlined the general principles of form. We here insert this axiom: Form is a fundamental property of existence; therefore, that which has no form has no existence. Limitation is a property of form. The universe has existence; therefore it has form, hence it has limitation. While the above axiom partakes somewhat of the syllogistic method, it will be noticed that the objectionable feature of the syllogism is expunged; namely, the premise is not an assumption.

Motion obtains in everything throughout cosmic form. Nothing exists without motion. The atoms of the rock are constantly changing place with all other atoms. There is circulation in the bar of steel. The angular crystal atoms of the diamond are in motion, and in their circulation and impingement they generate electro-magnetic substance of the most delicate attenuation. All these circulations are regular and according to the fixed laws of order; therefore, while form exists according to definite principles of form, the laws of motion conform to and determine the principles and arrangement of organic relation and shape.

Cause of Motion

The cause of alchemico-organic motion is remote and proximate. Before defining the laws and principles of motion, we will here briefly state that within the alchemico-organic world (cosmos) there resides the anthropostic or corresponding cosmos. These are two discrete spheres, yet they are co-ordinately one. The alchemico-organic cosmos (the physical world, the outline of which is given in preceding pages) is in the form of man; that is, in the egg or shell--man in the state of un-incubation.

The mass of humanity is in the same state, with this modification; the alchemico-organic cosmos is in space, and is therefore persistent; while its anthropostic co-ordinate embodies principles which merely correspond to space, and are not persistent in any one form. For instance: The seven metallic laminæ or plates comprising the general metallic rind of the macrocosmic shell are perpetual. These are the seven limitations of essences radiating from the stellar nucleus primarily, and from the solar limbus as the environment of the stellar nucleus. They constitute the deposit extremities of the seven essences, or the seven alchemico-organic spirits of radiation. The geologic strata through which the essences radiate are the conditions of chaos penetrated by the essences before they reach their extremities of metallic deposition and order.

The stellar nucleus is the center of space; the metallic laminæ are at the circumference of space. Correspondentially in humanity, the Lord Christ was the stellar Center, and his quality was the correspondent, in anthropostic being, of space in the alchemico-organic cosmos. In the progress of time in its relation to the development and progress of the race, the seven churches yet to be formulated into groups are the anthropostic depositions corresponding to the seven metallic plates. The seven churches are seven qualities of human characteristics and correspond to the seven planets, and therefore to the seven primary substratic laminæ of the cosmic crust.

By the remote cause of motion is meant the cause primary to the electro-magnetic substance created at and radiating from the stellar nucleus, antithetically generated at the circumference of the shell and converging to the nucleus. The cellular cosmos, or the great cosmic egg previously described, constitutes a great electro-magnetic battery which is purely physical, or, as denominated in Koreshan Science, alchemico-organic. The sun and stars are focalizations of physical spirit-substance, merging into matter materialized through voluminous and high-tension convergence.

There are at these centers constant concretion and sublimation. Spirit-substance is constantly materializing, and the temporary materialization is as rapidly changed to spirit-substance and is radiated. There is, therefore, a reciprocal interchange of substance from center and circumference. The spirit-substances engendered at the nucleus are radiated to the circumference, and are there solidified. At the circumference the surplus solidification is reduced again to spirit-substance and flows to the nucleus.

As there are seven metallic laminæ in the prime circumference, so there are seven prime metallic kinds of essence flowing toward and into the stellar nucleus. As these influxes are of seven distinct characteristic vibrations, so the nucleus has seven distinct degrees of essence, all meeting at one focal point in space, there turning back upon themselves and flowing out or radiating to the circumference and depositing at the environments of the cosmos. From the mineral laminæ, geologic strata, and water surface of the universal rind, there is also a corresponding and co-ordinate inflow or convergence to the stellar nucleus.

However, we find in this great universal battery, in its electro-magnetic power, but the proximate cause of its activity and form. Thus far there is no conscious and voluntary spirit-substance. Associated in co-operative being with this alchemico-organic cosmos is another half, endowed with voluntary and involuntary consciousness, co-existent and co-eternal. This voluntary and involuntary conscious existence, the acme of whose life is in the human brain, while prior and positive as to the momentum of this duplex cosmic structure, is only coincident as to its perpetuity. Neither existed prior to the other in the timic aspect of their co-ordination.

The proximate cause, then, of all the motions of the alchemico-organic cosmos is electro-magnetic substance produced reciprocally at the center and circumference of the great alchemico-organic battery, by the destruction of matter; for let it be reiterated and remembered, that spirit-substance is the result of the destruction of matter as matter; and that matter is the result of the destruction of spirit-substance as such. In other words, an atom of matter is the materialization of physical spirit-substance, and this substance is the dematerialization of matter. Both matter and spirit are substance. It will thus be seen that spirit-substance is not therefore a mere mode of motion or of vibration, but in reality is something in motion.

The Remote Cause of Physical Motion

We have hinted only, in a general way, at the proximate cause of the activities in the alchemico-organic cosmos. The term remote cause is here employed as being a cause remote from the electromagnetic essences upon which immediately depend the form, motions, and phenomena of the cosmic structure outlined in the foregoing. We are now to consider the very central and primary cause of all motion. This cause is mental. Not only is it mental, but it is voluntary and of the will.

There are two cosmic fields of form and function belonging to discrete degrees, but yet co-ordinate and interdependent. These are the alchemico-organic and the organo-vital. The first and lower is that which embraces the world as the earth, with the stars, sun, planets, atmosphere, etc.; the other, the higher, is the vegetable, animal, human, angelic, and God kingdoms. They are both co-eternal; neither existed before the other. The organo-vital is prior as to quality, and prior also as to its positive power to create and perpetuate.

The alchemico-organic field centers in the astral nucleus as the positive pole of its electro-magnetic essence; the organo-vital centers in the divine Man, the bright and morning Star, whence originate the voluntary redemptive will and creative power. The Lord Jesus is the representative nucleus of the regenerated manhood. The Lord Christ at the time of his manifestation was the center of the anthropostic universe, the source of being, the point and origin of creative power.

Cause of Motion From the Biblical and Theological Point of View

"He is the image of the invisible God, the first-born of every creature: for by him were all things created that are in heaven, and that are in earth, visible and invisible, whether they be thrones, or dominions, or principalities, or powers: all things were created by him, and for him: and he is before all things, and by the all things consist; and is the head of the body, the church." (Col. i:15-18.)

There can be but one question regarding the above Scriptural declaration. Is it true? It is concise; sweeping, inclusive, conclusive, and lucid. Is the Lord Jesus the Christ of God what he declares himself to be, and what inspired men have declared him? Is he the Son of God? And does He embody, as the primary offspring of Deity, all the attributes of the parent? And more than this; in His development from men as the Son of man, did he absorb into himself the principles, attributes, life, form, and consciousness of the parent?

We hold that the Lord, as was declared of him, was the fulness of the Godhead bodily,--Father, Son, and Holy Spirit; and further, that when the Lord was visibly manifest to the outer world, his inner and spiritual life was visible to the spiritual world as the astro-biological center of that sphere, and beside him there was no God.

How could the Lord, being born an infant into the natural world only at the beginning of the age, be the cause of all things? The Lord was not only the reincarnation of Elijah (God the Lord), of Moses, of Abraham, of Noah, of Enoch, and of Adam, in a direct line, but of all who died looking to his coming as the Messiah and Son of God in the indirect lines of reincarnation. He gathered into himself the spirits of the past. He was also the pole of influx from the heavenly worlds, and constituted the rolling together of the heavens as a scroll. He was the Word infolded and sealed. "Him hath God the Father sealed."

Twenty-four thousand years before the beginning of the Christian age, conditions in the world were the same as then. God was manifest in the flesh, and the Lord of the Christian era was identical with the Adamic personality in the beginning of the 24,000-year cycle. The end and beginning of every grand Zodiacal cycle bring into visible and personal manifestation the Son of God, who is Father, Mother, and Son.

He, the Lord, was the individual or undivided man. He held within himself the Bride, for "He who hath the bride is the bridegroom." He was the biune, the two-in-one, the parent of himself, and also of the Sons of God. As the parent of Himself, High Father, he was Abram; as the Father of the Sons of God, he was and is Abraham. He being the very primate cause of all things, and possessing both the voluntary and involuntary power of creative being, it is seen that the cause of all things resides in voluntary mental substance, supplemented by the involuntary reflex of voluntary mental activity.

Precession of the Equinoxes as Related to Astro-biological Manifestations

When the ordinary "scientist" alludes to the precession of the equinoxes, he has reference merely to the sun's precessional movement; but every planet passes through a corresponding precession. The sun's precessional year is 24,000 years long. The precessional years of the planets are correspondingly longer, proportioned to the difference in their ordinary years. These precessional years constitute cycles of time that are definite and recurrent, and proportionate to the great complex, solar, lunar, planetary, and stellar precessions.

The movements in the alchemic-organic sphere have an astro-biological correspondence. The signs in the physical heavens mark definitely the manifestations which correspond to them in the astro-biologic field. Every 24,000 years there is a similar personal manifestation as the one constituting the beginning of the Christian era. Every 24,000 years there is such a manifestation as is now about to occur.

We are now approaching a great biologic conflagration. Thousands of people will dematerialize, through a biologic electro-magnetic vibration. This will be brought about through the direction of one mind, the only one who has a knowledge of the law of this bio-alchemical transmutation.

The change will be accomplished through the formation of a biological battery, the laws of which are known only to one man. This man is Elijah the Prophet, ordained of God, the Shepherd of the Gentiles. and the central reincarnation of the ages. From this conflagration will spring the Sons of God, the biune offspring of the Lord Jesus, the Christ and Son of God.

The Transposition of Mental Force to Physical "Energy"

When a man (the man) so understands the laws of life as to know their application, and through obedience to law overcomes the sensual tendencies of, his nature, he reaches the point of biologic absorptions The visible and tangible dematerializes, and the outer consciousness enters into unity and blends with the inner and spiritual. The visible man consumes and thus enters, by transabsorption, into the unity of the invisible Godhead, and, by descent, into the church prepared to receive the precipitate afflatus.

In the process of the dissolution of the visible structure, by which the matter of the tangible organism is dissolved, consumed, and reduced to spiritual substance called the Holy Spirit,--the substances (oxygen, hydrogen, nitrogen, carbon, sulphur, phosphorus, fluorine, chlorine, sodium, calcium, potassium, magnesium, cuprum, aluminum, iron, etc.) contained in the organic structure, together with the atmosphere and the free physical spirit-substances of space, enter the vortex of vibration, which consumes the body.

The consciousness of the man entering thus into the whirlpool of organic dissolution is not obliterated, but infolds by conjunctive unity with the central and interior mind, around which the outer consciousness had wrapped itself. This interior mind constitutes the very central consciousness of Deity, the heart of the anthro-biologic cosmos. Radiating from this afferent absorption, the gravic spirit in its efferent distribution baptizes such minds as are prepared to receive the divine overshadowing, called the Holy Spirit.

Every overshadowing of the primary seven successive baptisms proceeds directly, not from an invisible "oversoul," but from the tangible personality. The Holy Spirit shed upon the world (the church) in the beginning of the age, proceeded from the visible Lord in his conversion from matter to spirit. If the Lord had not been personal, there could not have obtained the dissolution of his body and its conversion to Holy Spirit; and therefore the afflatus (the "oversoul") could not have obtained.

But the vortical involution, the anthropostic, took place in alchemico-organic (physical) space; therefore it involved the material elements not included in the organo-vital structure, and the vibrations were communicated to the elements and essences of the alchemico-organic cosmos, converging toward and into the stellar nucleus of the alchemico-organic world, and radiating to its circumferences.

It is thus that the conscious mental nucleus of the anthropostic imparts momentum to the activities of what has been denominated the physical universe. The Impulse ceases to be mental substance as soon as the influence of correlation has mutated the vibration of mental substance to the vibration of alchemico-organic (physical) motion.

We do not employ the term vibration as it is usually employed in common psychical or physical "science;" with us, vibration signifies not merely motion imparted to atoms or essences, but the dissolving of molecules and atoms, and their conversion to spirit-substance in the various degrees, and vice versa. It is thus observed that a reciprocal relation exists between the anthro-biologic (organo-vital) and the alchemico-organic worlds, and that the electromagnetic essences active in the latter are the result of a continuous, primary, voluntary mental essence generated in the human mind, dependent on the material basis (brains); these two co-ordinate fields of operation being co-eternal.

Cause of Motion of Planetary Disci

There are seven primary, movable, mercurial disci floating between the metallic plates or laminæ. The momentum is imparted through the operation of the actinic radiations from the astral nucleus, projected through the solar influence. These radiations penetrate the geologic, mineral, and metallic strata.

As the sun radiates its substance in the form of a cone, the apex of which is at the solar center, and the base at the metallic strata, the impression made upon the strata is in the form of a circular area. There is an impression of the alternate influence of caloric and cruosic substance, the one expansile, the other contractile. This movement follows the rotation of solar motion, therefore there is necessarily a peristaltic or vermicular motion imparted to the metallic plates or laminæ.

The alternate action of the actinic radiations of calorine and cruosine produces discular vacui between the plates, which are filled with mercurial amalgam. These act as reflectors, throwing back into the heavens the forms of the disci against the atmospheres, so that in looking toward the heavens we behold these disci through the operation of the laws of reflection, and are thus enabled to comprehend how the Lord "spreadeth out the heavens . . . as a molten looking-glass."

Knowledge of Universal Form Necessary

A knowledge of the structural form and function of the alchemico-organic (physical) cosmos is the key to our knowledge of the principles which must govern the organization of society in the culminating kingdom of righteousness. The importance of a knowledge of universal form and function, as pertaining to the alchemico-organic cosmos, will be admitted when the mind is sufficiently amplified to comprehend the relationship of the alchemico-organic macrocosm to the organo-vital macrocosm (the Grand Man), as pertaining to and comprising the universal mass of human existence.

The individual (undivided) man (such a man was the Lord) is the archetype of creation. What He was in the least form, the alchemico-organic world is in its greatest form; and what he was in that form, so also is the final

Theocratic Kingdom in the earth, namely, in the form and function of the man. Therefore, we discover that the true interpretation of the alchemico-organic cosmos is the revelation of the mysteries of Deity; for as the outward and most material structure is but the expressed thought of the voluntary and involuntary mental cause producing it, so a knowledge of this expressed and manifest language reveals the history of human origin and destiny.

OPTICAL FACTORS AND ILLUSIONS

The Proofs of Cellular Cosmogony Contrasted With Assumptions

IT IS ASSUMED by those who profess to believe in and advocate the Copernican system of astronomy, that the earth is convex because it appears so from optical observation. A person standing upon a tower and looking out in every direction will see the vanishing point at an equal distance, and the horizon (the limit of geolinear vision) describes a circle around this center of observation. This fact in appearance is taken as an assumption of the earth's convexity, because it is claimed that nothing but a globe would thus respond to and impress itself upon the organs of vision.

We maintain that an assumption predicated upon an optical illusion is not sufficient ground for the establishment of a rational conviction. If the earth were a perfectly flat surface extended illimitably, an observation from a tower looking out in every direction would assume, to the eye, the appearance of a circular horizon, for the simple reason that geolinear foreshortening would provide for a vanishing point at a given distance from the observer, proportionate to the elevation of the point from which the observation is taken.

If a person will stand upon a railroad track equidistant between two rails, the rails will seem to approach each other in the distance, the apparent contact, or vanishing point, being proportioned to the space between the rails and the height of observation... If they are five feet apart, the vanishing point is less than if they were six or seven feet apart.

Suppose we take a geolinear extense on the surface of the earth as one rail, and an imaginary line through the air as the other, placing the eye two and one half feet from the earth's surface. Now, the same law obtains in looking parallel along this surface, as in looking parallel to the rail and along its side. Making our observation by the side of the rail, the vanishing point is reached and the rail disappears, although extended in a straight line far beyond the vanishing point. The line over which observation is taken along the surface of the earth is the geolinear extense; it corresponds to the rail, and disappears by the same law; namely, that of foreshortening.

Deceptive Appearances

Appearances vs. Facts

The phenomenon of the disappearance of a ship, hull first, as it recedes from view, is caused by the same law of foreshortening as that which governs the disappearance of the rail, or causes the two rails to appear to approach each other. If we should make calculations on the basis of the appearance instead of on the basis of the fact that the rails do not approach, but only seem to, we necessarily draw false conclusions. This is precisely what the astronomers do. They conclude from appearances rather than from facts.

A balloon six or seven miles distant, appearing about the size of a pin head, if it be sixty feet in diameter, occupies as much space in the distance as when near the subjective point of observation. The law by which the balloon appears to diminish in size as it recedes from view, is the same as that which produces geolinear foreshortening, or which makes the surface of the earth diminish longitudinally as extending from the point of observation.

This phenomenon belongs to the organ of vision, and cannot be comprehended only as we possess a correct knowledge of the laws and phenomena of optics. Owing to this fact, the student cannot comprehend the principles involved in the phenomena of optical appearances and illusion without a thorough comprehension of the principles and laws of optics.

In another part of this volume the reader will find a complete record of the mechanical apparatus and processes by which we have so absolutely demonstrated the concavity of the earth as to over-shadow the fallacious conclusions of the mountebanks,--Copernicus, et al, and their deluded followers. We place a brief study of optics before the reader, merely to show wherein the fallacious conclusions of modern, so called science, while conflicting with the discovered and projected truth, are drawn not from facts but from appearances.

The Laws of Visual Impression

It might appear, as we proclaim the fact that a thorough knowledge of the Koreshan Cosmogony demands a thorough knowledge of optics, that it is our purpose to set forth a complete optical treatise preparatory to an understanding of the Koreshan Cosmogony. A thorough knowledge of Koreshanity must necessarily be a question of growth. A slight knowledge of the laws of optics will enable the student to see the discrepancies of modern astronomy, as predicated upon a misinterpretation of appearances.

What we behold through the organs of vision depends entirely upon the imprint of objectivities upon the retina of the eye. What we see is merely a picture placed upon the lining coat of the eyeball, and thence carried through the optic nerve, optic commissure, and optic tract, to that cortical area upon which the final function of vision depends.

We refer to a diagram setting forth some of the correlated facts of vision. The reader's attention is again called to the explicit study of the effects of subjective impression, or the imprint or picturing of the objective world upon the retinal coat. (See retinal coat in Diagram 1, Plate 1, with the area b b as the film upon which the imprint is laid.)

The picture upon the retina includes whatsoever is embraced in the obtense between the two lines 1, 2, 3, 4; a a a is the optical axis, d is the point of the appearance of the ship when the hull vanishes, as it recedes from view, as observed from the subjective x. The dotted lines indicate the appearance of the actual lines 1, 2, 3, 4, while d is the apparent position of the ship observed from x (the subjective point), and c, the ship as it actually is, viewed from its location in fact, not in appearance as at d. The perpendicular space 1, 1, implants the picture f h; the space 2, 2, implants the picture e g.

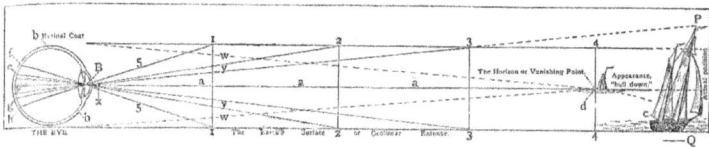

DIAGRAM No. 1. Illustrating "The Laws of Visual Impression."
This Diagram Illustrates a Principle, not Measurements True to Scale; the Height of the Objects Is not Proportioned to the Distance.

It will be noticed that the picture imprinted from 2, 2, at e g, is shorter than the one imprinted from 1, 1, at f h, proportionately as the distance from 1, 1, to 2, 2, in the objective. It follows that if a picture is imprinted from 3, 3, at b b, the ratio of shortening at b b will correspond to the imprints, 1, 1, and 2, 2. If lines were drawn from the points 4, 4 to the film b b through the focus at B, the subtense of the angle from 4, 4 to B would be so acute as to obliterate the space at the center of the film b b.

The point of obliteration at the film or retina, b b, corresponds to the vanishing point in the objective at d. At d the hull of the ship disappears, because there is no longer room for the picture upon the retina.

The lower line 1, 2, 3, 4 is the geolinear extense; the line upon the ground appearing at d, the vanishing point and the horizon. The upper line may represent a cloud covering the sky. The two points 4, 4, appear to join at d because of the distal foreshortening, which it must be remembered is merely the result of changes upon the retina, effected by distance. Any object beyond the ship c, as seen at d, will settle out of sight on the geolinear surface, proportionately to its distance beyond 4, 4.

By comparing the spaces w w with the spaces y y, it can readily be seen how the area of a given space appears to shorten, and narrows itself upon the retinal coat. Now if we remove the upper line 1, 2, 3, 4 and open up the space above, an object at P may imprint itself upon the retina; but an object at Q could not be seen because it is below the ground surface, which, though it might extend a thousand miles in a straight line, can make no further imprint upon the retina because the space between the lower line 4 and d is the obliterated space, as affecting the retinal film.

We have presented some optical facts upon which depend the appearances upon which rest the fallacies of the Copernican system; facts, a want of the understanding of which places the so called scientists in the catalogue of the incompetents, which graces the contradictory systems of astronomy that arise spontaneously, subserve their purpose, and die the death of the fallacious in the various careers of mental transformation, as the human mind gropes its way in darkness.

KORESHAN APPLICATION OF GEODESY

New Method of Determining Earth's Contour

GEODESY is the application of mechanical and other means for the purpose of determining measurements of the earth's surface, including not only that of its general contour as to whether it is concave, flat, or convex, but also of demonstrating the amount of curvature at any given point and in any given direction.

The Copernican system of astronomy assumes that the earth's surface is convex, and upon this assumption the fallacious system has been fabricated. No astronomer has ever yet presented any proof of the Copernican system; and one of the persistent efforts of the modern physicist is to find some irrefragable proof of what every so called astronomical scientist knows to be merely an assumption.

The Koreshan System of Astronomy is in direct opposition to the Copernican system, and unlike the Copernican system it is founded, not upon an assumption, but rather upon a premise so absolutely within the sphere of mechanical demonstration as to place it beyond and out of the uncertainty of mere postulation, which we assert to be the basis of so called, modern science.

Heretofore, the common method of attempting the determination of a straight horizontal line has been by the use of the engineer's level. There are a number of optical factors not taken into consideration by the geodetic surveyor and civil engineer, which render it impossible to extend a horizontal rectiline by the aid of optical instruments. The engineer's level is an instrument used by the surveyor, and includes a level and small telescope usually placed on the top of a tripod. This is more especially employed for the measurement of angles.

It is a fact not generally known, that it is impossible to determine a horizontal rectiline with a leveling instrument, or by the unaided eye, along the apex of successive heights of a given elevation, or along a continuously extended surface. The scientific reason for this impossibility resides in the fact that in the determination of a horizontal or lateral rectiline, an impression made upon the retina of the eye by a picture from one side of a visual direction must be counterbalanced by an equal picture on the opposite side; and the geodetic engineer, not being acquainted with this law of obtension in optics, extends a curved line while he believes he is continuing a rectiline.

Two men of different heights cannot, while adjusting the tripod to accommodate the difference, extend a line of the same curvature. A civil engineer six feet tall--adjusting his tripod to conform to his height--will make a curved line, by the aid of his instrument, upward of a given curvature, while the man five feet six inches tall, adjusting his tripod to suit his height, will determine the curvature of a lesser curve proportionably to the difference in height of the adjustment.

The scientific cause of this discrepancy resides in the optical illusion referred to above, namely, that on one side of the visual line there are two factors entering into the formation of a picture on the retina, as follows: The perpendicular post producing the effect of retinal impression is shortened or elongated proportionably to the distance of the object in perspective; and in addition to this, the geolinear foreshortening (the line along the earth's surface) induces a corresponding effect upon the retinal membrane.

We confront, then, two kinds of foreshortening--the one geolinear, the other perpendicular--in all geodetic observations; and an optical phenomenon which should be attributed to the principle of perspective foreshortening is ignorantly attributed to curvation.

To obviate the introduction of optical science and the necessity for the explanation of optical illusions and intricate phenomena incomprehensible to the ordinary mind, we have instituted a simple mechanical device by which a rectiline can be determined. (See diagram No. 2, Plate 1.)

DIAGRAM No. 2. The Rectilineator Used in the Koreshan Geodetic Survey.

The Rectilineator

Perpendicular standards are placed at points where there is a quiet expanse of water large enough in area to extend a line six, seven, or more miles. Across these perpendicular standards the horizontal bar of the Rectilineator is adjusted. From this first adjustment the rectiline is extended in both directions, until the line meets the water at a distance proportionate to the height of the perpendicular standard.

By this operation we extend a chord from the top of the uprights, at right angles to two points at the surface of the water, as in diagram No. 4, Plate 1. The relation of the straight line to the arc determines the concavity of the earth as its true contour.

In diagram No. 3, Plate 1, we have an illustration of the optical effect of an observation made with a leveling instrument, which does not differ in principle from a corresponding observation made with the unaided eye. The straight surface over which the line of observation extends is represented A A A; B B B is the visual direction deviating in a gradual curve away from the straight line A A A. The mind is unconscious of this curvation of vision, hence the curved line appears to be straight as in the dotted line C C C, while the straight line A A A appears to rise gradually as the line D D D.

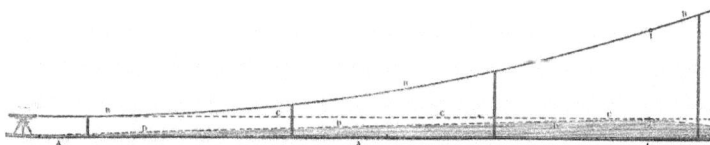

DIAGRAM No. 3. Illustrating the Illusions of Optical Phenomena.

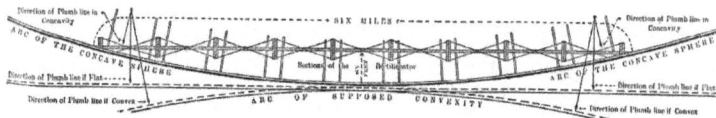

DIAGRAM No. 4. Comprehensive View of the Air Line, Showing Use of the Rectilineator in Survey of Chord of Arc by the Koreshan Geodetic Staff at Naples, Fla.

The point 1 in the line of vision appears to be at the point 2. The vanishing point is where. the extremity of the visual line at 1 seems to meet the line A A A, represented by the line D D D. Beyond this point the straight line A A A, appearing as the line D D D seems to convex away from the apparent line D D D. This optical phenomenon, which is an illusion, is taken as a demonstration of the convexity of the earth, and made the basis of the illusory system of the Copernican astronomy.

In the observation illustrated by diagram No. 3, Plate 1, we prove that a straight surface curves away from the line of vision, by the identical argument employed to prove the convexity of the earth. We can prove that a straight line bends four different ways, by the same argument used to sustain the convex theory of the earth.

Revolution in Geodetic Surveys

Revolution in astronomy implies revolution in all things. The great Swedish Seer said: "Every dispensation proceeds as from an egg." We reiterate, that a scientific religion which must embrace scientific social organization, will proceed from an astronomical basis, the foundation of which is the Cellular Cosmogony. Life develops in the cell. When the world is forced to accept this proposition, all else follows readily.

In connection with the establishment of the fact, in the public mind, of the concavity of the surface of the earth, and next also in importance is the determination of the amplitude of the arc, or the radius of its curvature. This cannot be determined accurately by any process of surface triangulation, because there are too many factors entering into the process to insure accuracy.

The Rectilineator, extending its line from an given height of a prime vertical, approaches the normal curve of the surface at a proportionate ratio, which may be determined at any given point by two exact methods,-- each acting as the verificator of the other. Place a perpendicular at the requisite height, about six feet, more or less, and adjust the initial section of the rectilineal bar at right angles.

The points selected should be as nearly level as possible. After the extension of the line three or four miles (even less than this will answer), adjust the geodetic level. This is an instrument having two graduated glass perpendiculars very minutely spaced, with microscopes adjusted to the graduated side of the glass tubes. These two perpendicular graduates are united by a connecting tube twelve or fourteen feet long. (The tube and graduates contain mercury.)

The amount of variation of the mercury in the graduates, with the connecting tube arranged parallel with the rectiline of the section bars at any point, will indicate the degree of curvature. The instrument must be perfect; this accomplished, the determination of the radius of curvation is most simple.

This instrument may be verified by the use of another instrument adjusted to the section bars with a perpendicular rod, to which is adjusted a very slender plumbline. Across the bottom of the rod, which has a flat surface, is a minutely divided scale, to which is, also adjusted a microscope. The scale has a definite number of divisions to the inch. This will determine the amount of variation from the prime vertical; namely, the first perpendicular.

The deviation from the normal will increase either from the prime vertical, as the line extends, or toward it, according to the direction of curvature.

This method of mensuration determines both the direction of the curve and the radius of curvature. Any portion of the surface of the earth can be a thousand-fold more accurately surveyed by this method, than by any process ever instituted. We know that the result will compel the world to acknowledge the Koreshan System of Cosmogony.

LAWS WHICH DETERMINE THE FORM AND FUNC-
TION OF THE UNIVERSE

STUDENTS of Koreshan Universology are becoming familiar with the term geodesy, and the phrase, the new geodesy; and that there may be no misunderstanding of its meaning and its bearing upon life, we think it admissible to define its significance and its relation to the system of universal culture which Koreshan Universology embraces. The term geodesy is from two Greek words; ge, earth, and dai, to divide. It is literally the science of dividing the earth, or of defining its character as to form, that there may be a foundation for a knowledge of its functions.

The Sloop Ada.
In service of the Koreshan Geodetic Expedition on the Gulf of Mexico.

There are three fundamental laws involved which, when understood, determine the form and function of the universe as an entirety. These are first, the science of comparative cellology,--the foundation principle of which analogically determines the fact that all life, whether that life be specific or general, unfolds within the cell. The law and principle of comparative evolution analogically determine the fact. that universal life is a unity, and that the progress of gestative evolution must necessarily progress within the great cell or womb of creative incrementation; second, the science of vision, known by the term optics, in which are interpreted the appearances of objects on the surface of the earth as related to the earth's contour.

At this point let us quote a passage of Scripture which has a very significant application to the subject under discussion: "Judge not according to the appearance, but judge righteous judgment." This law applies as well to physical observations as to moral, religious, and spiritual things. Things are not as they appear at all times; hence the necessity for understanding principles, that interpretations may be genuinely true.

Book optics and practically applied optics are two very different things. The reader must become familiar with practically applied optics. For instance, books will tell you that because the earth is convex, three posts placed in the water three miles apart will conform to the convexity; the middle one will be the highest of the three on the convex surface; and looking from the initial post toward the terminal one, the line of vision will cut the middle post and strike the terminal or distal post at a higher point than the middle one. No one pretends to dispute this fact of observation.

Perspective Foreshortening an Essential Factor

The interpretation given and generally believed is, that the world is convex; and because it is convex, and vision being in a straight line, the fact is according to the appearance. Place three posts three miles apart, the distal one being six miles from the initial post--the three posts being each one foot above the water's surface. Now place the eye, unaided by the telescope, at the top of the initial post and look toward the middle and terminal post. The middle and distal posts will be out of sight, not from the fact of convexity, but from the fact of perspective foreshortening. Place a telescope of about three inches diameter of the objective lens, upon the initial post; you look over the top of the middle post and see the distal one on a curve above the middle post.

The truth concerning the matter is, that vision is deceptive unless the science of perspective fore, shortening is applied to the interpretation of the first and second observation, the one with the unaided eye, and the other with the aid of the telescope. The fact that the books and practical experiment do not agree should serve as a precaution against believing all the books say, when those books contain only theory founded upon assumption.

The science of optics, then, may be called the second science applicable to geodetic discrimination, and one of the laws employed to corroborate the testimony of comparative cellology, which determines the contour of the surface of the earth, and the fact that the earth is a great electro-magnetic cell. It should be remembered, that comparative cellology settles the question of the concavity of the earth, and the fact that man in-habits the earth. The science of optics corroborates the testimony of cellology.

The third science is that of mechanics as applied to the measure of the contour. It will be noticed that there are three methods of proving the fact of the concavity of the earth's surface. The first and greatest is comparative cellology; the second and most complicated, the application of optics; the third and most simple, by mechanical application. In the perfection of a treatise on the new geodesy these three principles would necessarily be included, for the reason that the corroborative testimony of more witnesses than one is essential to conviction of different characters of mentality.

Geodesy Applied to Earth Measurement

Geodesy is the application of mechanical and other means for the purpose of determining measurements of the earth's surface, including not only that of its general contour as to whether it is concave, flat, or convex, but also of demonstrating the amount of curvation at any given point and in any given direction.

The Copernican system of astronomy assumes that the earth's surface is convex, and upon this assumption the fallacious system has been fabricated. No astronomer has ever yet presented any proof of the Copernican system; and one of the persistent efforts of the modern physicist is to find some irrefragable proof of what every so called astronomical scientist knows to be merely an assumption.

Our knowledge of the figure of the earth is only obtained by indirect means.--Astronomer Ball.

The geodetic operations carried on during the last century and a half for the purpose of determining the figure and dimensions of the earth have, up to this time, led to no satisfactory results. They have been performed by the most eminent astronomers, with the most perfect instruments, and it would seem that they ought to have led to a final solution of the problem; such, however, is by no means the case. Every new measure of a meridian arc has but added, and adds, to the existing doubts and want of concordance, nay, to the positive contradiction which the various operations exhibit, as compared with one another.--Von Schubert.

The Koreshan System of Astronomy is in direct opposition to the Copernican system, and unlike th Copernican system it is founded, not upon an assumption, but rather upon a premise so absolutely within the sphere of mechanical demonstration as to place it beyond and out of the uncertainty of mere postulation, which we assert to be the basis of so called modern science.

UNIVERSAL FORM AND FUNCTION ARE PERSISTENT

WE HAVE SHOWN the sphere to comprise the outlines and limitation of universal form. All convergent lines from the circumferences of the sphere determine toward and terminate in the center, which must necessarily constitute the focal point of centri-petal flow--the central point of contact of all material things. The physical universe being the formulated expression of mind, the astral center of the physical universe must comprise the analogical correspondent of the astral center of the mind of the universe.

The physical universe is proximately moved by the essences of alchemical, electrical, and magnetic action, which are generated by the very form and relation of the elements which enter into the constitution of the great cell or shell of generation. Remotely, it is moved by the mental potencies that are above and prior in quality to even the unconscious or material forces; such as physical lumen, calorine, electricity, magnetism, levity, and gravity.

Universal form and its correlate universal function are persistent, never having had beginning only so far as the modifications of time (by the break in continuity) mark the end and the beginning of periods, and denote timic aspect, or that modification of continuity called time. "In the beginning God created," does not imply more than the beginning of a specific cycle; and such creation, or recreation, is, manifest whenever a cycle closes in the beginning of a succeeding one.

If function and form are correlate and persistent properties of perpetual being, (and there is a sense in which things were not created,) still the continuity of the universe must depend upon its recreation or creation in an existent form and function. That is what is meant by creation.

Atheists, infidels, materialists, and spiritists may continue to rave and rant at the obduracy of man's adherence to what they may denominate a "book of fables," it yet remains as an enduring monument of a never-fading glory of conception as far beyond the intelligence of the ranters as they are below, in descent, an ancestry from which they maintain they have degenerated. We mean, of course, the monkey, chimpanzee, and gorilla, of which modern atheism seems so proud to boast for ancestral origin.

After thousands of years the Bible remains the bulwark and citadel of towering strength, unaffected by the onslaughts of its enemies. Religious systems, founded upon false interpretations of the Scriptures, may rise and fall, but the truths of the Bible never--they are eternal.

The Universe Is One Vast System

Koreshanity regards the universe as one vast system, with such a perfect adjustment of parts as to embrace every department in a combination of co-operative unity and procedure, not merely as pertaining to the solar and stellar realms, embracing the earth inhabited by man, but inclusive of mineral arrangements, and vegetable, animal, and human life, Man, both as to his individual and universal being, is the archetype of the cosmic structure and function. He is both origin and product of the integral co-ordination of universal mode and motion.

The mind of man, inhabiting his organic form, is the positive pole of constructive potency, and the material cosmos has proceeded from his voluntary purpose and co-ordinate involuntary consociation. The laws of construction, with the potential and kinetic essences through which the depositions of circumferences are formulated and maintained, and primary and subsidiary centers pivoted, focalized, supplied, and regulated, correspond to the laws of organic unity consociating mind and body.

The universe is the great ovum of integral incubation. In-cube-ation is the modification of the lines and forms of the cube and sphere, adjusted to the purposes of use in the integral economy. The chick is incubated (hatched) from the egg, the infant is incubated in the matrix. Humanity as a whole is incubated within the great cell or ovum of universal life and not (contrary to the universal law of development) on the outside of an uneconomic adjustment and compilation of matter, as men throughout the world of boasted civilization have been blindly taught.

The forms of the incubated are but the modified adjustments of curvilinear and rectilinear motions and modes; or, as applied to man, he is the cell doubled upon itself, and the laws of mental and organic life correspond, in him, to the laws of motion and arrangement in the functions and emplacements of the physical cosmos. The unique and complex cell, with its correlation of circumference and center, is the expressed or evoluted form of integral being, and the forms and laws of this comical integrality are the pattern of the integral government of man.

Discovery of Koreshan Universology

IN THE WINTER of 1870 I discovered the science of the universe and coined the word to suit it; namely, Universology. The first time I ever saw the word was after I had coined it, by compounding it of Latin and Greek words, the first part being Latin, and the last part the Greek logos, word or discourse. I do not say that the word might not have been previously framed and employed. I do not know when Stephen Pearl Andrews first employed the term; but I do know that I was not conscious of any use of it until after my application of the word to my science.

In the discovery of universal science (I might say in the revelation to me of what, up to that time, had been a mystery to the world at large) I found the universe to be a composite whole, an integralism having been in perpetual existence in the past, and must be. in perpetual existence in the future. I discovered the earth to be stationary, a macrocosmic and composite shell or rind, circumferential to the atmospheres, and the solar, lunar, and stellar manifestations within this common rind or composite and concave sphere.

The universe is a fixity from more points of view than one. It is creation from God who dwells in it, and derives his functional powers through his inherent relativity to the integralism of the universe of which he is a part-- an intellectual part, individual and recurrently personal. The formation of the universe is discovered to be cellular and therefore concave, not upon the basis of hypothesis, but upon the actual and positive demonstration by three systems of geometrical calculation, including optics and mechanics.

When the "scientific" world and the world generally discard the speculative and hypothetical methods of arriving at conclusions, and begin to reason from demonstrated rather than from hypothetical premises, they will begin to awake to some startling facts, and to realize the astounding absurdity of any and all conclusions predicated upon the guesses of men ignorant of the first principles of the origin of life and the destiny of the human mind.

The most advanced thinkers of the times are seeking the origin of life. This is ignorance seeking light but finding none, because the keys of knowledge have been taken from the world by the doctors of the world, under the influence of scholastic and classic cult. No wonder that the question is being agitated in the scholastic circles! The subject could not be promulgated for years from the Guiding Star Publishing House without awakening interest and discussion.

The Perfect Man Is the Archetype of the Physical Universe

The description of creation given in the firsts chapter of Genesis was never intended primarily as a presentment of the physical cosmogony, but as the creation of the man himself, which in his most external manifestation is in the physical form. Thus in this outward degree the heavens comprise the various regions of the mind, while the body constitutes the' earth. This, however, applies only to the integral man, and not to the so called humanity who have not attained, through the restoration, to the image and likeness of the Gods. While the genesis of man is described in the first chapter of the Hebrew Book, it is also true that the perfectly structured and integral man is the archetype and microcosm of the physical universe; and when the comprehension of the one is reached, then also is that of the other.

The physical sun is the center of the physical universe, and the earth is its crust, shell, pediment, or rind. As the sun is composed of light and heat as its primary substances, so the mind correspondingly is composed of the affectional (love, or desire) substance, and its accompanying wisdom and intellection. These two principles and substances are the primal substances of being; the one is the heat, the other is the light of the mind; and these two substances extend and exert their influences in the body to effect and perpetuate its substance and form, with all the accompanying phenomena.

The activities of the mind, so far as they relate to the organic structure of the body, proceed from the conjunction and unity of the two primary activities and substances of mind. These substances and activities eventuate in the formate solidarity of the organic natural structure, through the formation of another substantial force; namely, the substance of gravity. Weight substance, or as it may be technically called, gravic substance, is the first product of the union of light and heat, and is the first law of form.

As one of the fundamental principles of knowledge, we here reiterate that the law of gravity is the first law of formation or creation, and is the product of the union of light and heat; and that the body is the product of love and wisdom, the two corresponding substances of mind.

KORESHAN PRINCIPLES OF OPTICS

IN CONSIDERING the objections against the present astronomical hypothesis about to engage the reader's attention, we would first invite a consideration of the principles of optics as enunciated through the pages of Koreshan literature, and especially to the fact that visual sub-stance generated in the gray matter of the cerebrum and cerebellum has more to do with the function of sight than the motion of extraneous physical force. Therefore, while studying this objection, the reader will remember that in our presentation of the inconsistencies of the Copernican system, we are also presenting the modern astronomer's views of the action of physical substance.

Vision, according to the present conception, radiates from any given center of the substance called light, and penetrates the pupil of the eye, making its impression upon the retina, whence the impression is conveyed through the optic nerve, commissure, and. tract, to the cortical area of vision in the gray matter of the cerebrum.

Every radiation of substance from the sun or any of the stars enters our atmosphere at some angle of deflection; the angle of refraction being proportionate to the divergence of the ray from the central or vertical one. In observing any star except the star vertical to the point of observation, it is observed through an angle of refraction at the point where the substance enters the atmosphere.

No living being--supposing for the sake of the argument that a ray of light penetrated our atmosphere from the sun or stars--could possibly determine the angle of refraction without knowing the distance of the limit of the atmosphere from the point of observation. The depth of the atmosphere is conceded by all astronomers to be only approximately determined; and no two astronomers are agreed as to the atmospheric depth.

Any man can positively know that the angle of refraction cannot be determined without a knowledge of the ray of incidence, and that if the angle of refraction is not known, the direction of a star observed through the ray cannot be determined.

It is positively known that no angle of refraction of any given ray of light from an objective source beyond the atmosphere, can be determined unless the exact depth of the atmosphere is absolutely known to the fraction of an inch.

In observing a star, either with the unaided eye or with a telescope, at an objective point divergent from the vertical direction, if the atmosphere were forty-five miles in depth, to the fraction of an inch, the angle of refraction could be determined, had we an exact knowledge of the difference in the tenuity of the atmosphere and of the ether beyond.

If we could determine the angle of refraction we could determine the direction. If the atmosphere were just ninety miles in depth, as some astronomers affirm, then from that knowledge--were it absolute--we could determine the direction.

Problems That Demand Explanation

If the atmosphere were forty-five miles in depth, the amplitude of the arc of its curvation could be accurately determined and the degree of refraction equally known. The depth of the atmosphere, whether forty-five, ninety, or five hundred miles deep, must be positively known before the amplitude of its arc can be known, and before the amount of radiatory deflection can be determined.

We know that astronomers, in making observations, pay no attention to any refraction at the supposed summit of the atmosphere. We also know that in works on physics and civil engineering, it is claimed that allowance is made for what is supposed to be the refraction of the atmosphere.

If the atmosphere refracts three inches to the mile, in an observation made along a horizontal line, how much does it refract at any given direction from the horizontal to the vertical? If observation is made of a star at or near the horizon, and the first mile shows a deviation of three inches, what will be the amount of deviation at any uncertain distance of forty-five, ninety, or five hundred miles? All of these estimates have been made by various observers and calculators.

If the earth curvates eight inches to the mile, it is estimated, on a calculation made at a ratio inversely to the square of the distance, that at the distance of three miles the deviation from the direction of the optical tangent is about seventy-two inches, or six feet. If the atmosphere refracts an optical line three inches to the mile, then at the distance of three miles-- by the application of the same law--the second mile would be nine inches, and the third mile twenty-seven inches.

If the third mile affords a deviation of twenty-seven inches from the tangent of the rectiline, making the calculation upon the basis of the inverse ratio of the square of the distance to the uncertain limitation, which may be forty-five, ninety, or five hundred miles, (as yet undetermined by any of the astronomers,) what will be the amount of deviation at the unknown and uncertain point--the limitation of the atmosphere? This is the question to which the Koreshan Cosmogonist demands an answer, and to which the investigating world also demands an answer.

These considerations entering as factors into the problems of astronomy, demand some explanations regarding the fact that the astronomers do come to correct conclusions. Upon the basis of the ordinary calculation of the earth's curvation, or even a simple divergence of an optical line from a rectiline, supposing the divergence at the objective end of a telescope twenty feet in length to be only an eighty-one millionth of an inch, what would be the direction and location of a star trillions of miles distant?

KORESHAN COSMOGONY IN CONTRAST WITH MODERN ASTRONOMY

WE HAVE SHOWN the true character of cosmogonic form, and have placed this revelation in contrast with the uncertain Copernican system of astronomy. We have devoted much energy and effort to bring the questions of Koreshan Universology prominently before the people for public discussion. In this effort we have been held up to insolent ridicule and most bitter persecution, consonant with the invariable rule to which every innovation upon prevailing public sentiment is subject. We would not be worthy of consideration if our doctrines were not important enough to excite the animosity of the sentiment in both the secular and religious phases of thought which our system assails.

We have pushed our claims to a knowledge of cosmology until the advocates of the spurious "sciences" begin to feel their insecurity, and the necessity for defending their right to the title of "scientist" and "scientific." So long as the "scientific" world rested in absolute security upon the ignorance of the laity, it felt no necessity for the discussion of the question of the Koreshan Cosmogony; but our persistence in the advocacy of the truth, in contrast with the audacious assumptions of the Copernican advocates, incites a growing uneasiness regarding the stability of an astronomy which has nothing but assumption upon which to rest its claims to acceptation.

The whole batch of assumption and absurdity called modern science is assaulted by the consistent and determined purpose of the apostles of Koreshan Universology. We know that when our system is considered of enough consequence to receive candid notice from thinking men, and when the advocates of the prevailing system of astronomy begin to comprehend the fact that their premises, which they confess to be mere assumptions, are being analyzed by honest investigators, and are known to be worthless as foundations for the building of the superstructure of science, they will be compelled to make an open defense of their untenable position.

The Copernican system of astronomy had its rise in the dark age; and there is not an astronomer of note who does not know and confess that there is nothing but assumption for its foundation. It is responsible for the agnosticism so much in evidence, and for the attitude of that stupendous farce, the "higher criticism." There is not a phenomenon manifest that is not easily and rationally explained and accounted for from the standpoint of Koreshan Universology, whether belonging to the domain of physical or psychical manifestation; and per contra, there are no phenomena, either psychic or physical, rationally accounted for on the basis of the modern system of so called science.

Darwinism Is a Fair Sample of Modern Scientific Conclusions

Says Darwin in "Animals and Plants," (Vol. I, page 9): "In scientific investigations it is permitted to invent any hypothesis, and if it explains various large and independent classes of facts, it rises to the rank of a well-grounded theory."

It is to this absurd proposition that most of our "scientific" theories, if not all of them, owe their existence. He further says, that "The undulations of ether and even its existence are hypothetical, yet every one now admits the undulatory theory of light."

We agree with Darwin that the undulatory theory of light is a mere hypothesis; that is, a mere guess; but we deny his statement that "every one now admits the undulatory theory of light"

Darwinism, as Darwin himself affirms, is predicated entirely upon "scientific" guesses; and these, he declares, constitute the basis of all scientific claims. Speaking of natural selection, he says: "Now, this hypothesis may be tested,--and this seems to me to be the only fair and legitimate manner of considering the whole question,--by trying whether it explains several large and independent classes of facts; such as the geological succession of organic beings, their distribution in past and present times, and their mutual affinities and homologies. If the principle of natural selection does explain these and other large bodies of facts, it ought to be received."

"Please accept my theories," says the eminent "scientist," "because I can explain many things upon my hypothesis." The Koreshan scientists might beg the question and say, please accept our theory of Universology, because there is not one thing that we cannot explain scientifically upon our premise. But we ask no man to accept anything on the basis of a mere hypothesis. A knowledge of the construction of the universe and its functions, with the laws and principles of life depending upon such knowledge, is too important a matter to be left to mere conjecture--mere hypothesis.

No conclusion is certain which is not founded upon and grounded in a positively demonstrated premise. It is for this reason that the Koreshan System stands out distinct and unique. It predicates nothing upon guesswork; its first step in the discussion of any proposition is the correct establishment and proof of its premise. Darwinism is a fair sample of the processes by which modern scientific conclusions are invariably reached.

Facts and Appearances Differentiated

Let us take the principle of optics in its application to the definition of the phenomenon of the rotundity of the earth, as an illustration of correct reasoning from an established premise, as followed by the logician of the Koreshan School of Science. We herewith accompany our argument with diagrammatic illustrations of the principles involved in the argument.

Two lines may be extended parallel with each other, as in the case of the two rails of a railroad track. The diagrams represent certain known facts in optics, which we declare shall not be overruled, set aside, nor ignored for the purpose of sustaining an unwarrantable "scientific" theory. If any man is too lazy to reason, or too mean to investigate another's reasoning, we do not expect to make any impression; or if he is so wedded to a theory because his grandfather believed in it that he will not change his opinion for the truth's sake, he will naturally pass this argument by; but for the honest man there is only one alternative.

The two lines, a f, extending the length of diagram 1, may be taken to represent the tracks of a railroad, five feet apart. In the major premise of this proposition are involved the facts as they are, not as they appear. The measurement of the space at both ends of the track shows that the rails at each end are just five feet apart. There is no element of assumption in this part of the premise. We wish to thoroughly impress upon the student the fact, that so far we have not had to "assume something." The rails are straight and parallel, and five feet apart. These are facts of practical and certain measurement.

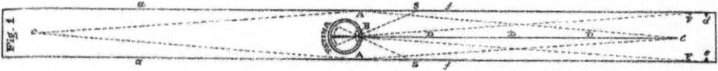

DIAGRAM No. 1. Illustrating the Vanishing Point of Space Between
Parallel Railway Tracks.

From B to c in either direction the track indicates one mile; (the entire length of the diagram representing two miles;) in observing the distance from B to c, either way, the track appears to narrow down to a vanishing point at c. This appearance is the minor premise. Let it be remembered that the minor premise involves a fact, but that fact is an appearance, involved in which are certain optical laws which we will apply logically in another part of this argument.

Do not forget the fact that we are arguing from premises that are proven to be true, and that we differ from the ordinary "scientific" logician, in that we work from a demonstrated premise--not from an assumption.

The purpose of this part of the argument is to show the reason for an appearance which is in direct opposition to the fact. Why does the space five feet wide at F F appear as a point at c? Note the dotted lines beginning at S S and extending to the arrow; they make a comparatively long picture upon the surface indicated by the arrow: Note the dotted lines beginning at F F and extending to the surface marked retina. These lines vanish at the point upon the surface thus marked; for this reason they appear to come to a point at c.

If we take this appearance as a fact, we are led into an interminable labyrinth of difficulties. The "scientist" establishes his assumptionsupon these appearances, ignoring the facts and laws of optics. The objects e d are in fact at e d, as represented in diagram 1, but they appear to be at c.

We are to distinguish the facts of reality from the facts of appearance, and show the character of the appearances, and how these appearances have led the short-sighted "scientists" into their aggregate of errors, which they delight to call by the title of science. Thus far there is no element of assumption; we assume nothing.

The Law of Foreshortening Ignored by Modern Astronomers

We have shown that space is annihilated in appearance by the law of distal perspective; that foreshortening is an inevitable law of optics, and we hold that these laws are totally ignored by every so called scientific astronomer. The pseudo scientists shall not continue to foist their fallacious systems of astronomy upon a deluded public without a perpetual protest.

It will be remembered that diagram 1 represents the point of observation at B, from which the objective point is seen at c, but which in reality is at F F. The line D D D, extending to c, is not what it appears to be from the outlook or visual point at B. The apparent line at c; which appears to be only a line, is the entire breadth of five feet--the distance across the track at F F. If a middle rail extend midway between the two rails of the track it will be seen the entire length of the line, or nearly so, and seem to blend with the two other rails at c; the five feet have vanished to a mere point at c, therefore a space five feet wide appears like a mere line.

The broader the space in perspective, the more rapidly it vanishes by distal extense, as shown in comparison with the middle rail; and the narrower the space, the less rapidly it vanishes by distal extense. This principle belongs more exclusively to the effect on the retina itself. A balloon in passing out of visual range appears to diminish rapidly for the first few miles, after which it remains in view for a long time as a mere speck. These facts will have their application during the course of this argument.

We subjoin a second diagram. Here we have two lines as in diagram 1, but we will employ them to represent parallel lines, one above the other instead of side by side, as in the first instance. The line A A appears to rise to B B, and the line C C appears to drop to B B, if viewed from the point D. The points A A and C C are visible, but they are seen as if at B B. Now, is there any man capable of thinking, who will be such an obstinate ass as to take this appearance as the fact, after the phenomenon has been pointed out to him?

Principles of Optics Applied to Geodetic Observation

We have studied the phenomena of appearance in these principles of optics, and will now proceed to make an application of them to geodetic observation, keeping logically to the premise, never swerving from the established law of Koreshanity; namely, that assumption is no basis for the establishment of truth.

We subjoin the third diagram. In this we take the lower line of the second diagram, A A; we observe the points A A from the point D, but the principle of perspective or distal foreshortening causes the objects to appear at the points B B. This is not due to refraction, but it is due to distal foreshortening; the space from A to K has contracted and foreshortened to the point B. This law is operative and applies to all space whether in the atmosphere, ethereal, or on the surface of the earth, terrestrial.

If the line A A in diagram 3 represented a flat surface, a convex surface, or a concave surface, the phenomenon would be practically the same; a convexity or concavity of only eight inches to the mile would not appreciably affect the optical illusion. If the so called scientist is asked the question, Why does the earth viewed from a balloon look like a bowl? he will tell you it is because of atmospheric refraction.

DIAGRAM No. 2. Showing Same Perspective Effect With One
Rail Above the Other.

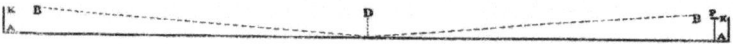

DIAGRAM No. 3. Showing the Vanishing Point or Horizon of Geolinear
Surface, With Upper Rail Removed.

If the laws of refraction will operate in an atmosphere of uniform density to distort the vision, what may we not expect regarding phenomena related to objects claimed to be outside our atmosphere? If the point D, in diagram 3, is two feet above the line A A, at the distance of less than a mile the object at A on the lower line will be seen at B, in either direction. The law is the same whether the altitude be two feet and a half, five hundred feet, three miles, or any distance. A less or greater altitude could not change the principle nor alter the character of the phenomenon.

The cross-piece at P is seen at A, but appears to be at B, because the standard, A P, is foreshortened down the two feet and a half. We have thus far shown certain facts, and optical phenomena connected with these facts. We have assumed nothing regarding the facts or the phenomena. We have interpreted the phenomena by defining the laws upon which they depend, and we challenge all the scientific men in the world to point out one inaccuracy either in the facts as presented and pertaining to the reality of the relation of the lines, or the facts of the optical phenomena.

Our Minor Proposition

We are now prepared to state a minor proposition. Lines or surfaces separated by narrow or broad spaces (extended parallel with each other and viewed in perspective) will appear to approach each other proportionate to their distance from each other and length of perspective. Let the surface of the earth be taken as one of these surfaces, and extend a line over this surface; that is, a visual or optical line. If we stand twenty feet above the surface of the earth and look toward the horizon, the horizon is seen on a level with the eye.

If a roof could be extended parallel to the surface of the earth twenty feet above our lookout (forty feet above the earth), the two surfaces would appear to approach each other; the lower surface would seem to rise to a level with the eye, and the upper surface would appear to drop to a level with the eye;--that is, providing the two planes are extended the necessary distance.

If we remove the upper surface or plane, the lower plane will appear to rise to a level with the eye, just the same as when the upper plane occupied its position. It would not be occasioned either by refraction or convexity, but would be due to the operation of the principle of foreshortening. How a man can observe this phenomenon and attribute it to anything but its true cause, and call himself scientific, is one of the enigmas of this so called enlightened age.

We have practically shown that the apparent rotundity or convexity of the earth is due to the optical illusion created by foreshortening. When it is assumed that the earth is convex, and in this assumption the simplest laws of optics are set aside and ignored, shall we quietly submit to the imposition and allow the world to continue in ignorance of the laws of cosmogonic form, or shall we place the facts in opposition to the assumptions as they obtain and are made to constitute the basis of scientific conclusions?

If a man stands by the side of one of the rails of a railroad track, say two feet from the rail, his line of vision will meet the rail at a point determined by the distance in perspective. This we need only state, for it is a well-known fact. No man will pretend to deny this, unless he be an absolute ignoramus. Then why should anyone deny the phenomenon as applied to the extense of any other line or plane?

If the earth were concave, eight inches to the mile, which would be a practical level and an apparent straight plane, and we should apply the law of optics as described, in looking along a geolinear surface the earth would appear to rise in perspective much more rapidly than the eight inches to the mile would indicate. If we were twenty feet above the surface of the earth, the earth would rise to meet the line of vision and would appear to be convex.

The scientific and honest man, before he projects a theory on the basis of appearance, would submit the appearance to a rigid analysis; he would prove his premise by the facts, and not ignore the most common principles and laws of optics as applied to geodesy. Let us demonstrate our premise, then reason logically, and we are certain of the truth. Let us assume our premise from mere appearance and then make our theory fit the premise, and we have just what the scientific world is attempting to cram down the throats of the credulous and unthinking public.

Our Sub-major Proposition

Our sub-major proposition is, that a rigid mathematical calculation, founded upon the mathematical determination of the amount of foreshortening of the space between any two given parallel lines or planes in perspective, when applied to the surface of the earth, will determine the amount and direction of deviation which the surface of the earth describes, from a line extended from the point of observation to the vanishing point.

Let two lines be separated by a definite space, and extended parallel to a distance sufficient to obliterate the space by distal foreshortening. Extend these lines one mile parallel, a definite space intervening, then apply the same distance in length with the same space to any other two lines, and the same results would obtain. The truth of this statement is obvious to any candid person.

If we make an observation along the side of a line which we suppose to deviate a few inches one way or the other from a rectiline, and calculate they difference between the definite foreshortening of the known lines and space, and the space of the indefinite line, the difference is the amount of deviation of the unknown line. This will also determine its direction.

Let this principle be applied to the surface of the earth, and the demonstration will determine whether the earth is flat, convex, or concave; also the amount of deviation, if any, from a plane. The claim that the earth is convex is made upon the mere appearance from optical effects, without any consideration of the laws of foreshortening, and the whole system of cosmogony is made to fit this absurdity.

We have pursued this argument from a known fact, and have applied a knowledge of the laws of optics as related to these facts, to the appearance of the surface of the earth as under the operation of these laws. We have shown that the laws of optics prevail and operate in the one case as in the other. We have shown that the laws of optics are totally ignored by the "scientist" in his consideration, and that he attributes an appearance to the application of an imaginary and impossible operation. It is also demonstrated in this discussion, that the principle of refraction is used as an argument by the so called scientist, where the principle of refraction does not enter into the proposition.

We have accounted for the appearance of rotundity on the basis of the known principles of foreshortening in perspective, which every sensible and conscientious man will admit to be obviously true. We have shown, then, that if the earth were an extended plane for ten thousand miles, a view from any altitude would give the earth the appearance of a rotund form, in dimension proportionate to the altitude; the greater the altitude, the larger the appearance.

We have shown what every honest "scientist" admits; namely, that the whole system of the Copernican astronomy is predicated upon an assumption which has no tenable foundation; therefore we are justified in our challenge of the accuracy of the system of astronomy which now flourishes under the title of "science." We also assure our readers that the time has come which the eminent astronomer, Professor Woodhouse, of Cambridge, England, feared would meet the so called astronomical profession. He said:

"However perfect our theory may appear, in our estimation, and however satisfactorily the Newtonian hypothesis may seem to account for all celestial phenomena, yet we are compelled to admit the astounding truth, that if our premise be disputed and our facts challenged, the whole range of astronomy does not contain the proofs of its own accuracy. Startling as this announcement may appear, it is nevertheless true; and astronomy would indeed be helpless, were it not for the implied approval of those whose authority is considered a guarantee of its truth. Should this sole refuge fail us, all our arguments, all our observations, all our boasted accuracy would be useless, and the whole science of modern astronomy must fall to the ground."

We have shown that the principles of optics have been left entirely out of consideration in the establishment of the Copernican hypothesis, and that therefore it is not worth one thought as constituting a basis for cosmogonic conviction.

The Koreshan General Proposition

We now state our general proposition: The astronomers of note admit that the whole fabric of hypothesis called astronomy is built upon an assumed premise of appearance. When a premise is assumed, the conclusion is necessarily an assumption. It is easy to fit a large aggregation of facts to any hypothesis; but this does not prove any proposition. An "hypothesis does not rise to the rank of a well-grounded theory," and never can so long as that hypothesis is predicated upon a premise that is itself not proven.

The earth is of some definite form; this form is absolute, but it has never been fixed in the mind of the thinker, for the reason that, up to the present time, the "scientific" world is looking for some positive proof of the earth's rotundity, its revolution on its axis, and its orbital motion. These have never, so far, been regarded as settled facts. Because of this uncertainty we claim the right to demand some better reasons than have ever been adduced for the acceptance of the Copernican system of astronomy, and an examination into the reasons we have promulgated for a disavowal of present "scientific" claims.

AN ABSOLUTELY INDISPUTABLE PREMISE

THAT the perpendicular pole forms a right angle in both directions with a chord, the extremities of which meet the verge or horizon, is given as the first fact,--supported by the statement of every scientist, and corroborated by thousands of experiments. When a man views the horizon he does so along what is called a horizontal line, which is always at right angles with a perpendicular one. Now let us compare the above facts with the diagrams and theory given by the advocates of the accepted theory of cosmical form.

No. 1 is the usual diagram employed as the first step in the demonstration of the convex rotundity of the surface of the earth. The subjective point, the point of vision, is at A. The horizontal line, or line of vision, is at B, and C is at the objective point. The subjective point usually represents a man, and the objective point a ship in the distance. Let the reader carefully note the relative angles of the line representing the man at A, and the horizontal line B.

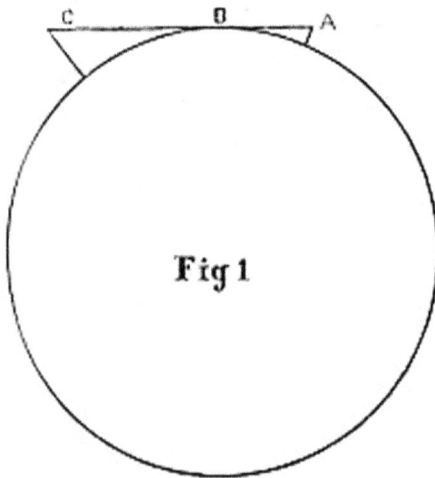

Fig 1

It will be observed that the man at A stands obliquely to the horizontal. If this relation of the two lines is compared with the facts as actually observed in natural phenomena, there is discovered a disagreement. Man stands perpendicular to the earth, and at right angles to the horizontal line. No scientific man living can reconcile this disagreement with the commonly accepted cosmical theory.

In diagram 2 is shown a continuation of the horizontal line C B to G. The vertical line A forms an acute angle with the horizontal line C, but an obtuse one with its extension B G. These angles as represented in the diagram are contrary to facts as observed in Nature.

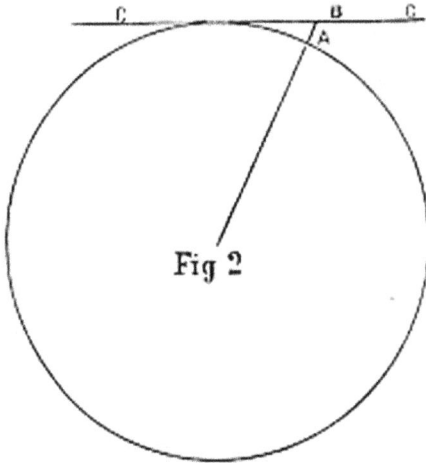

Fig 2

The physicist has but one escape from the dilemma he has gotten himself into, and that is the denial of the fact of the horizontal direction of vision toward the point where the earth and sky, or water and sky, seem to meet.

Let the reader hold the book in such a position as to give the axis A B, diagram 3, the vertical direction,--A up and B down. C will be a horizontal line, forming an acute angle with D, which represents the point and relation usually given in the diagrams presented to school children, as a rudimental step in their study of cosmogony. While C is a horizontal line, and D an oblique one, and while a spirit level would indicate the horizontal toward C, the spirit level would indicate a declining line toward E. This is not true according to fact.

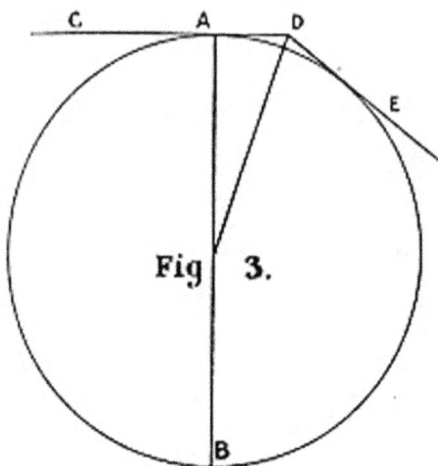

Fig 3.

If a man place his point of vision ten, twenty, thirty, forty, fifty, or a hundred feet above the surface of the earth, and it be unobstructed by natural or artificial interferences, he can observe the horizon on a level with his eye.

A, in diagram 4, represents the visual point, B B the line of vision in both directions, C C the points indicating the verge at horizon, D the base of his position resting upon the earth, which. describes a curve downward from C to C. The straight line, C C, forms a chord, and the curved line, C D C, the arc of the chord.

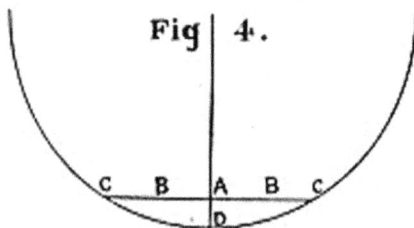

Fig 4.

From certain cognized and indisputable, collected factors we have formulated a premise as absolutely indisputable, upon which we establish the great and cardinal scientlflc truth of Koreshanity; namely, the concavity of the earth's surface. The first factor is, that a man standing plumb with the "center of gravity" (base of gravity, which is on the circumference of the sphere, and center of levity, which is at the center of the sphere) maintains a perpendicular or vertical relation to the surface of the earth.

The second factor is, that a line drawn from the eye, or point of vision, at any given distance from the surface of the earth, in opposite directions. from the visual center, touches the verge or horizon on a level with the visual center, and that the chord thus described from horizon point to horizon point is at right angles with the perpendicular line maintained by the vertical posture of the man. These are absolute facts, easily verified by any person who may take the pains to inquire into the physical phenomena. These facts are in direct contradiction to the cosmological theory of modern science.

The Koreshan Premise

The premise, then, from which we demonstrate the concavity of the surface of the earth may be stated as follows: A horizontal line drawn in opposite directions from any visual center touches the earth's horizon at the two extremities of the chord, and the arc of the chord forms a depression from the center of the chord, equal to the depth of the perpendicular radius-vector.

The extension of the curve necessarily completes the circle of the earth, which comprises the circumferential sphere of the solar system. The astral center, or central star, is at the nucleus of this sphere, around it being the luminous sphere comprising what forms the sun proper, from which proceeds the projected sun at the limit of our atmosphere.

Between the earth's concave surface and the solar sphere there are three atmospheres. The first one is composed of oxygen and nitrogen; the second one of hydrogen, and the third one of aboron. These atmospheres occupy the first dimension in space. Occupying the same space but comprising a second dimension, is a series of spheres composed of physical spirit located at seven distinct distances between the astral center and the circumference of the earth.

The earth constitutes a circumference, the focus of which is the astral center. The diameter of this circumference is about eight thousand miles. The distance, therefore, from the center to the circumference is four thousand miles. From the center there constantly flow toward the circumference the physical spirit-substances generated within the sun, or at the astral center.

These spirit-substances: flow toward the circumference, and are met by co-ordinating spirit-substances flowing toward the center from the circumference. At the point where the outflowing and inflowing substances meet, a new substance is generated from the action of the two, which comprises the potency of revolution. (This sphere of spirit-substance is one of the spheres already noted above.)

The earth's crust or shell is composed of seven metallic layers and five mineral or earth deposits. The location of the metallic strata may be determined by taking common atmosphere as the zero point, water as the ratio, and the given specific gravity of any one of the metals as indicating the point of location of the aggregate and static sphere of the metal. The metallic layers form a compound pile or battery, of which the voltaic pile answers as a sort of representative.

Between these spheres, that is, between each pair of metallic strata, there is generated a spirit-substance which flows toward the center. There are as many kinds of spirit thus generated, as there are spaces or conjunctives between the layers. These seven qualities of spirit-substance meet as many outflowing substances, and at the points of meeting in space produce seven spheres. These are the planetary spheres, the planets being the focal points. There are seven metallic planes, from which are focalized seven planets in the physical heavens.

The Law of Visual Deviation Accounts for Certain Phenomena

Thousands of objections will be urged outside of the argument thus far instituted, against our cosmological theory; but such objections, not coming within the logical steps of our argument already taken, do not demand any reply. We have opposed an insurmountable argument so far, founded upon indisputable factors. We need not, therefore, urge further demonstration of our Cosmogony until these objections are overcome.

We will, however, answer the very common objection in the mind of almost every person not willing to accept the Koreshan theory of Cosmo-gony. The objector urges the fact that a ship seen approaching in the distance, first presents the top-mast to the perception of the observer. If the old cosmogony were true, that is, if the earth were convex, the point of observation would be vertical to the center of gravity, this being at the center of the earth.

Staff Headquarters, Operating Station

Such being the case, the vertical point A, represented in diagram 5, would be perpendicular to the earth. This describes and locates the center of observation, which is the subjective point of vision; B B, objective point or top-mast; and C C declining, not horizontal lines. The letters C C are located where the lines touch the circumference, but these would not be horizon points, because not on a line level with the center of observation.

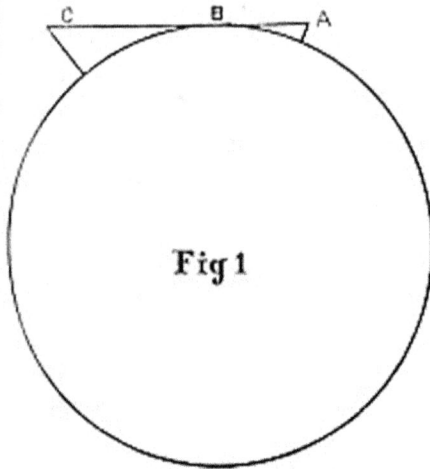

Fig 1

Any reasonable person can see that diagram 5 (barring exaggerations of diagram) would be the correct description of facts if the convex theory be the true one. School children should be presented with this diagram instead of the one usually employed. The diagram cannot be used because the line of vision is horizontal, looking toward and observing the horizon point, and because the deflection of the B C A, A C B lines, as shown in the diagram, is not true to Nature.

How, then, shall we account for the phenomenon which has so long deceived the "scientist;" namely, the observation of the top-mast of an approaching ship, on an apparently horizontal line from the subjective visual point? The law of visual deviation, which determines upward curvilineation of the visual line, accounts for the deceptive phenomenon, and settles the question of the concavity of the surface of the earth.

Diagram 6 represents the earth as a concave sphere. A represents a man standing vertical to the center of the sphere. The horizontal line of vision extends to E, the point where the line of direct vision touches the curve of the earth called the horizon. From this point the visual line curves upward. The visual curve is decidedly marked at this point. The curve of the concave earth is designated by the letter D, and the visual curve by E E. The theoretical and deceptive visual line is designated by F, and the theoretical and deceptive earth curve by G. The line A F is the apparent, but not the real line of vision. The top-mast of the ship appears to be in a straight line, for the obvious reasons given below.

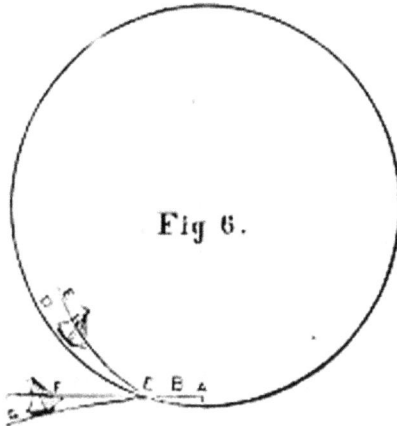

Fig 6.

A Commonly Accepted Law of Optics

It is a commonly accepted law of optics, that any observed object is seen apparently in a line corresponding to the direction of the ray entering the eye; as, for instance, let the vision be directed toward a surface of water, shown in fig. 7. A A is the water surface, B D, the line of vision, broken at C by the reflecting power of the water. D is an object in the air, but apparently seen at E, in a straight line from B. The perception is not conscious of the reflection. The rational faculty has therefore to be applied, to reconcile the facts with observation, and reach the real truth regarding the phenomenon.

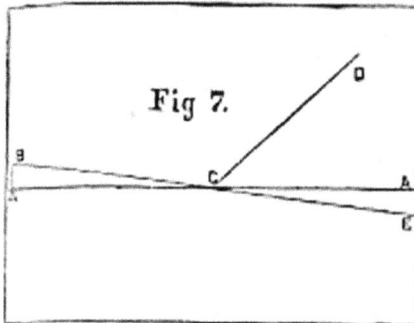

Fig 7.

If by any law there be a curvilineation of the visual line so as to make it deviate from the earth's curve upward toward the top-mast of the ship, the vision would be deceived in proportion to the amount of the curve from the horizontal or straight line, because the mere perception could not appreciate the curve; this appreciation belonging solely to the office of the reason as founded upon the facts of optics.

There is an upward curve of the visual line when perception is directed horizontally. This curvilineation depends upon a number of factors in optical law. The elucidation of this part of the subject involves a study of optics, as specially applicable to our Cosmogony. Vision does not depend solely upon the entrance of light into the eye from without. It is produced by the action of extraneous spirit-substance entering the globe of the eye through the pupil and lens, merging to a focal point in the center of the posterior chamber, whence it is radiated in all directions.

This substance stimulates the retinal coating of the globe, acting upon the retina (the expanded portion of the optic nerve), whence the impression is conveyed over the optic nerve and tract to the portion of the cortical substance of the brain upon which the substance of vision depends.

The cells of the visual cortex are stimulated, whereby they transmit visual substance back to the eye, through the optic tract and nerve, focalizing at the center of the globe, again radiating in all directions. The rays passing toward and through the double convex lens are brought to a focus just outside of the eye. From this focal point they diverge in every direction, passing out from the eye and touching objective points wherever there exists an obstructive point to reflex, or send back the impression of an object.

The rays passing out from the focal point act as telegraphic wires, so to speak, to carry back to the point of vision the return flow of substance, by which objective things are rendered visible by the impression these objects make upon the cortex of the brain. The real cause of visual curvilineation, technically stated, is refraction of gravity. It is the reaction of visual substance with gravic substance.

Cellular Cosmogony Dispels Idea of Illimitability

One of the first objections arising in the mind to the Koreshan Cosmogony is the apparent impossibility of sun, moon, stars, planets, etc., being limited to so small a space as the area of a sphere, the diameter of which is . only about 8,000 miles. Our system being true, the circumference of the sphere is about 25,000 miles, and its diameter 8,000. This, of course, would be an impossibility if these objects had the dimensions usually ascribed to them. The science of Koreshan Cosmogony dispels this hallucination, bringing the mind back to its rational conception of physical form.

According to the Mosaic description of creation God made two great lights; the greater light to rule the day, and the lesser light to rule the night; he made the stars also, and set them in the firmament of heaven to give light upon the earth, and it was so. Koreshan Cosmogony fixes the astral center at about 4,000 miles from the circumference of the solar system, this circumference being the earth.

Around the astral center is the solar sphere, which we call the sun. Outside of this sun are three atmospheres; aboron, hydrogen, and our common atmosphere, composed of oxygen and nitrogen. These three atmospheres extend from the sun to the circumference; namely, the earth, filling the entire space; and within these three atmospheres are the planets, stars, moon, etc.

The stars are the focal points of physical spirit, produced by its reciprocal reflection and refraction, flowing from the astral center through the solar sphere which surrounds it. The focalization of the stellar (star) points is produced by two systems of radiation and convergence, by which the transmitted "energies" of the sun, by virtue of the activity of the astral (star) center, are broken and converged to stellar points.

Material creation is the outmost expression of the thought of God. The Creator projects into outermost form and function only that which obtains in the divine mind, and that which he expresses represents the divine character and purpose when correctly interpreted.

A false translation of cosmical form, which is the expressed form of both God's desire (will) and wisdom, and the manifest phenomena of that form, is the basis of a fallacious theology; for man's conception and comprehension of Deity must agree with his interpretation of God's manifest expression in the physical universe, which is the unfolded cosmical speech or language of the Creator.

Astronomy is the law of astral or stellar motion and relation, and the concept we entertain of the physical universe, which is God's expression of himself, must correspondentially be the concept we entertain of Deity.

The sun is supposed to be the great center of the solar or sun's system. The emplacement of the "heavenly bodies," according to the modern physicist, is supposed to depend upon axillary and orbital revolution, and centripetal and centrifugal energy [so called]; that is., motion toward and from the center. If axillary and orbital revolutions are law of emplacement; in other words, if every heavenly body depends for its maintenance in its position upon the two motions, the one upon its axis, and the other upon its revolution in an orbit, then no center, no matter how aggregate the universe depending upon it and reciprocally related to it, could maintain its emplacement without both axillary and orbital motion.

SOLVING AN INCALCULABLE PROBLEM

The Easiest Way to Find a Solution Is to Jump at a Conclusion

OF ALL the absurdities of the Koreshan System, the world thinks the cosmogonical is the most senseless. "How can people entertain the belief that we are on the inside of a shell, contrary to the known truth that the earth is convex?" The fact that man occupies the inside of a spherical environ does not essentially resolve itself to a matter of contro-versy, for the statement of the truth carries conviction to every open receptacle of its fluxion.

Our promulgation of the cellular theory of the universe is not instituted without due consideration of all that it involves. We are prepared not only to show the contradictions, absurdities, and impossibilities of the Coperni-can theory of astronomy, but to meet every argument that may be adduced against our own, and to conclusively demonstrate the correctness of the Koreshan System.

In the original conception of the so called Copernican system of astronomy, in order to provide for the rapid passage through space of the occupying and moving worlds,--without their destruction by the friction that even the most attenuated substance would effect,--space was declared to be vacuate, as the most ethereal and attenuate substance imaginable would destroy, first, the atmosphere, then the water, and finally the substance of the solid earth of which the worlds (?) are composed.

It was subsequently discovered that as light and other [so called] energies were but the vibrations of ether, and that a vacuum was impervious to the transmission of "energy" of every description, space must be filled with something to provide for their communication. Here was a study for the mathematician. A body flying through space at the enormous velocity of 640,000,000 miles in three hundred and sixty-five days, besides the additional motion of 25,000 miles in twenty-four hours, must necessarily be free from the slightest encumbrance, hence the vacuum theory.

Now it is found that the vacuum theory will not work because "energy" cannot be transmitted through vacant space. How to calculate an incalcuable problem was the rub. Somebody scratched his head in profound thought, and here is the result. As we must have a vacuum to provide against friction, and substance to fill space to provide for the transmission of "energy," we must supply space with an imaginary something just thick enough for the "energy" business.

This must be purely imaginary, for otherwise we would have the friction. It must also be thin enough to obviate friction; and as the easiest way to find a solution to an incalculable problem is to jump at a conclusion, the difference was a compromise between the two extremes, namely, impalpability and nothing, and the great problem-solver split the difference between these two points. It was a great achievement for astronomy, and one of the first steps toward the overthrow of what the world, up to that time, thought to be a revelation from God.

Cannot any one see clearly, in the solution of the above problem, at the gait at which the great philosopher of attenuate ether notoriety conquered the obstacle, at what a rate the revelation theories of creation must fall into discount? What a mighty tumble for God and his visionary theories of creation; what humiliation for Moses, the medium of their communication to a benighted world, and what an opportunity for the lights of agnosticism to "don't-know" things out of existence!

The "don't know" theory--beginning at the point of attenuated ethereal solution for the accommodation of champion guessers, and the Godless Copernican basis of atheism, and ending with the unknown and unknowable god of modern Christianity--may answer for the ignorant who have not yet cast off the mantle of darkness adopted in the mediæval age; but for those who are emerging from the bliss of ignorance, nothing less positive than the perfect solution of the problem of life can afford satisfaction. The Koreshan Cosmogony will thrive because it is true. Its adherents are multiplying, and are the most intelligent of men and women.

Our sun, which would comprise the center of the solar system, with its planets, moon, etc., would revolve, with thousands of other similar solar systems, around another far distant center. All these solar systems, with their grand center, would revolve around still another, and this would in like manner depend upon a still greater one. This process of multiplication of centers and augmentation of the general system would continue without limit, no final center ever being reached. The physical universe would be without a definitely fixed pivot. A spiritual system fabulated upon such an astronomical system would necessarily leave out of the question a central and personal mind as the governor of the universe, hence the atheistic origin of thought, and atheism as a belief.

THE EARTH IS A SHELL; THE SUN IS THE CENTER

IN THE Koreshan System of Cosmogony the fact must always be borne in mind that the sun is the center, and the earth is the circumferential shell or environment. The sun has three primary or first principles; namely, heat, light, and gravity. These primaries are all complex. Heat is composed of degrees, light of spectra, and gravity of qualities. These [so called] forces are substantial in character, and are simply the most attenuate solutions of material substances.

Heat, light, and gravity are eliminated or emitted from the sun, passing through atmospheres which modify them according to the atmospheric qualities through which they pass. Every atmosphere has something of the power of reflection and refraction, but not so much influence upon the gravic substance as upon heat and light.

Solid metals refract and reflect gravity; every quality of gravic substance being refracted by its corresponding metal more than by any other, though they all have refracting and reflecting power over all the gravic qualities of substance.

The earth is a shell having seven primary metals in layers or strata, laminæ, planes. These constitute the rind or outer crust, and act as so many great reflectors and refractors of the [so called] forces. The direct reflection of heat is cold. The direct reflection of light is darkness. The direct reflection of gravity is levity. We have, therefore, cold, darkness, and levity, which are just as much substance as the three opposite ones. At the points of change are the poles of these substances.

I have emphasized the word direct, because indirect reflection partakes less of the nature of the opposite character, as it is less direct or more indirect. These reflex substances flow back to the center as cold, darkness, and levity, and move toward the sun in circular strata, taking their courses according to order, 1 between the qualities of the outflowing substance.

To give an idea of these return flows, let us take the dark force [so called]. In the examination of spectra there will be noticed the dark lines. These have been termed Fraunhofer's lines. In gravic analysis corresponding levic lines would be seen, and in heat analysis corresponding cold lines would be observed. These return flows are positive to the sun but negative to the eye. The outflowing substances from the sun are negative to the sun but positive to the eye. Thus the light is apparent while the darkness is not.

Origin of the Moon and of Gravity

As these [so called] forces flow into the sun they move toward its center in a vortex, which at the center becomes so rapid that they all commingle in a homogeneous fusion. The vortex produces, a cross circle which spreads out into a thin diaphragm, dividing the sun into two halves; one anterior and the other posterior. The central part of the sun (backward) is dark because the most intense dark "energy" seeks that point, while the central part (forward) is light. The dark "energy" moves out in the opposite direction from the light "energy." The cold "energy" moves out at one side, and the heat "energy" at the other; that is, at their most intense points.

The sun, then, has a double revolution--vertical and lateral. The vertical revolution is comparatively slow; the lateral is very rapid at the diaphragm, but less so at the back and front, or anterior and posterior points, which give to the mass the shape of two spirals or twists. Examine the heart and you will get something of an idea, as the heart represents one of the vertical halves. The diaphragm is one half-wheel cold substance, the other half-wheel heat substance. These substances spread out like two great wings, which extend into terminal levic rings that revolve from north to south, but not so rapidly as at the vortex and diaphragm.

The gravic substance is formative. It is more subtile and diffusive in its reflex than the other substances, and consequently more filtrative and general in its counter-flow or inflow, as it also is in its outflow, than the others; namely, light and heat. Electricity, magnetism, etc., are modifications of gravity and levity; but as these are let down toward the circumference through the atmospheres, they are directed by the power of reflection and refraction of the atmospheres. I have partially described the central sun or star center of the physical universe. We do not see the center, but only the focalization at the outer atmosphere.

To return to the diaphragm. The lateral rings, which are the peripheries of the cold and heat substances, assume the form of rings at that special relation, because a condensed ring of levity is the polaric opposite of thin laminæ, plates, or scales of gravity. The levic substance is not reformed from the mass until it is condensed at that periphery. This periphery is the aggregate positive pole of gravity, the aggregate reflex circle from the earth, and is there-fore the origin of the moon. By this you may see the truth of the Bible statement--"round tires like the moon."

What is the origin of gravity? It is the child of cold and heat. Cold is its father, and heat is its mother. The union of these produces the levic pole, the starting point of gravity, which solidifies in metallic form at the circumference--the earth.

Cause of Day and Night; also the Four Seasons

The sun in the third atmosphere is pendant, so to speak, from the pivot at the center. As the center revolves in its vertical revolution, the pendant sun in the third atmosphere moves in an orbit through the space of that atmosphere. This causes day and night. The lateral revolution in the sun, producing the diaphragm and peripheral rings (zones) of levic "force," produces the slow revolution of the cold and heat poles of the lateral cycle from north to south. From the cold pole of the zone (to midway between these extremities) it grows warmer, and from the hot pole it grows colder until the temperate is reached.

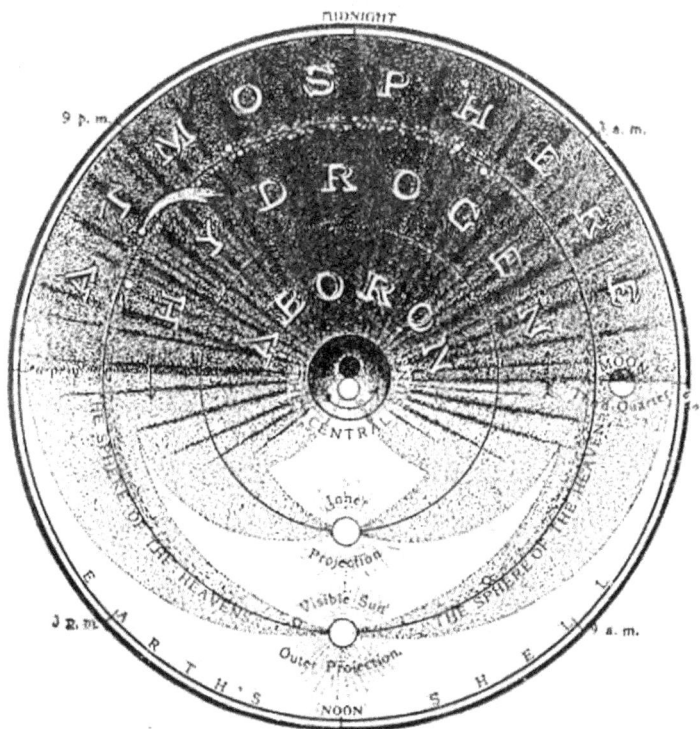

Cross-Sectional View of the Great Electro-Magnetic Battery, with the Sun as the Perpetual Pivot and Pole. Southern Hemisphere of the Cell

Here are four poles, the four winds of heaven, the four foundations of the earth, the four seasons. Now as heat and cold are the parents of levity, therefore, levity is at its maximum when the heat pole is south and the cold pole north; for then there is the greatest degree of heat at the south, and the greatest degree of cold at the north; but while the levic substance is at its maximum, it is not in its equilibrium in the zone, for contraction by cold diminishes it north, but increases it south. When the temperate poles are north and south there is a minimum degree of levic substance, but equilibrium in the zone, because the levic substance is equally distributed throughout the zone.

On the line of the ecliptic--which is the center or median line of the Zodiac--is a point of continual greatest gravity. This is a circle around the circumference, the point where the metals are thickest. This is not the equatorial line of the earth. The sun naturally pends toward this line; I mean the reflected sun in the third space or outward atmosphere. As he moves toward this line or ecliptic from north to south, or from south to north, he acquires a momentum induced by the attraction of gravity but regulated by the zone of levity.

As the sun moves north, levic substance increases gradually at that point, because expansion progresses in that direction by the action of the heat of the sum The expansion increases as the hot pole of the zone moves toward the north pole. When levic substance there is at its maximum, the point is reached of its greatest attraction for the gravic cycle in the ecliptic, and that side of the zone is pulled down, and the sun gets an impetus again toward the south.

As the sun moves toward the south, the levic zone increases at the south until the hot and cold poles are south and north, and the levic substance at the south, or hot pole of the zone, is at its maximum, when the attraction becomes sufficiently great again to overcome the momentum of the sun, and he takes his course again toward the north. These dippings of the rings are the moon's nodes or noddings.

These four poles, with the modifications of heat and cold in the zone, govern the sun's motion, and make the four seasons by the regulation of the sun's motion. It will be noticed that in the Koreshan System the ascending and descending nodes of the moon are accounted for, but on entirely different principles from the old theory.

The moon that we see is the transposed expansion (by reflection) of the levic zone, and the law governing the changes in the levic zone produces the moon's changes. The moon, so to speak, is the ghost of the earth, for the reason that the gravic substance takes the picture of the earth, transforms it into a ring or zone, and then again spreads the picture out against one of the atmospheres to our vision.

The stars, as distinct from the planetary system, are reflections from the central sun, back of the diaphragm, originating in the reflex flow from the diaphragm or from the vortical center. In the formation of the zone, at the extremity of the wings before mentioned, we have the commencement of the projected gravity. This does not extend in a continuous line outward to the circumferences or planes of the earth, but it is met by an interflow of levity, which is the point of polarity of the second degree.

The union of gravity and levity produces light, which from this second zone points forward convergently and forms an iris. This is the sun's anterior limb. The iris or rainbow is the circle around the pupil of the sun as it looks forward toward the earth. This is not seen in ordinary and uninterrupted vision; but when the spectrum from the sun is refracted by prismatic influence, this fact becomes apparent in prismatic phenomena.

In giving this anterior presentment I do so that the basis or foundation may be grounded for the efficient study of the posterior limb. The first zone mentioned is the zone of gravity. It is the first limbus, the limbus of kaneinic substance; that is, where it ceases to be in its greatest magnitude.

Meeting Point of Gravity and Levity

Gravity and levity meet at their first or highest conjunction, anterior to this zone. I have previously stated that levity was the direct reflection of gravity. Hence, at the point anterior to this zone, where the gravic and levic substances meet, at the rim or zone where the iris begins to converge, there is formed a posterior sheath or extension, constituting, in its direction toward the posterior limb, an elevated of nearly circumambient arch of levity.

This arch of levic substance is interrupted at points (cycles, rather) by effluxes backward from the posteriorly directed center from the vortex, and loses itself finally at a rim where the dark limb of the sun, in a posterior protrusion, forms a hole directly backward. This hole nearly corresponds to the optic nerve of the eye. It is formed by the reflex of light, and is what I have before termed scotos.

As the dark substance passes out, posteriorly, at the place of union or interruption of the circumambient sheath or dome, there is a gyre or spiral motion somewhat corresponding to the whirl at the vortex. This gyre loses itself, or merges into a fan-like protuberance. The posterior or crown is the preponderance of scotoic substance, but is intermingled with the levic substance, and also the irisic reflection. This crown is crossed up and down with striated bands.

These light bands, varying in color somewhat like the rainbow, (except that they are up and down, and do not form an iris like the rainbow,) correspond, in the dark posterior of the sun, to Fraunhofer's circular lines in the sun's anterior. I mean that when light is analyzed the dark lines are observed; but when darkness is analyzed the colored bands are seen. These bands are sometimes reflected in such a manner as to be seen in the heavens near the sun, and are called sun dogs. They no doubt derive their name from being projections from the vortex backward, as limbs of the kaneinic substance.

The bands of scotoic force are broad, while the stria are extremely thin or narrow, and stream out like so many colored fans. I call them stria, because they seem like furrows or deep grooves in the foldings of the cruosic substance.

These colored stripes nearly encircle the central orb from east to west; and as they radiate they strike an atmosphere, and, being refracted through it, point at substances in the formation of stars. These are the fixed stars. These colored bands do not form parallel lines like parallels of latitude, but are related to one another like degrees of longitude. The iris of the eye of the cat derives form from this factor in physical construction, although it partakes of both principles; that found in the anterior as well as in the posterior of the orb.

Cold the direct Reflection of Heat

We will now consider the cruosic or crystallic (cold) substance. This has its inception in the primitive zone, as I will now describe. In a former statement I defined the zone as having four polaric points, one of which is the caloric (heat) center or pole. It will be remembered that cruos (cold) is the direct reflection of caloris (heat).

Remember also that the zone has two movements; one from north to south, the other from east to west with the sun. The caloric pole, therefore, moves from north to south--as do all the poles. This movement from north to south, with the motion of the zone with the sun, accomplishes a caloristice corresponding to the summer and winter solstices.

The caloristitial movement is a circumpolar revolution of the caloric pole of the zone. The substance of that pole accumulates during the motion from north to south to the equator. This is gradually given off as it moves toward the south from the equator, but when it reaches its caloristitial circle, it is at the rim of its polaric aptitude; and there the heat is thrown in abundance to meet its polaric opposite. The direct reflection of this superabundance of heat is intense cold, not reflected immediately, but generated as the caloric pole of the zone moves again north toward the equator, and away from the circumpolar rim.

The caloric pole accumulates in potency again as it moves north toward the equator, after which it gradually yields its heat until the north rim is reached, where, in its circumpolar revolution, the caloristice, it throws off its superabundant substance, which is reflected as cruos (cold). Cruos is the substance of the Aurora Borealis, and is also seen as the corona around the sun during an eclipse.

The zone I have described is complemented by a second co-ordinating zone. The wings spreading out are lifted upward, so that this zone is above or backward from the diaphragm. The complementing zone is downward or forward. The relations of the zones, which are zones of complementary substances, are so related by the production of other substances as to form, in the completion of all their formations, relations, and unions, the cube and sphere, as also the larger zones and larger spheres, the whole of which comprises the real or true physical firmament, as fixed and solid in its structure as flint or steel; and the stars in these zones would be like diamonds in rings of steel.

These interior zones extend into space, where new zones and spheres are produced, governed by the laws regulating these interior substances, modified by the action of the earth-generated substances.

The Focal Point of all Influx. Planets Are Spheres of Physical Spirit. How Eclipses Are Produced

The astral or star center within the solar sphere is the focal point or center of all influx. All substances of every kingdom in the universe, generated as, the product of so called waste, as in the mineral, vegetable, and animal kingdoms, flow by graded progress toward and finally into this center. The various planetary spheres are stopping places, both from circumferences to center, and from center to circumferences; that is, spheres for the accumulation of substances. These are heaped up during the movements of planets in the aphelion, part of what is denominated their orbits.

The substance is thrown off or inter-transposed at perihelion or at the points of conjunction. These points of transposition are the centers of momentum to the physical universe. The planets are spheres of physical spirit having four primary focal points, each of these points representing the four kinds of substance in question, but each center being dominant in the manifestation of its distinctive quality.

The four qualities above referred to are mainly photos (light), scotos (dark), caloris (heat), and cruos (cold) . The spheres are arranged as shells around the astral center, between it and the general circumference, the earth. There are six of these spheres. While they comprise spheres located at six distinct distances, they are not spheres of solid substance like the earth.

When two or more focal points come in line, through the order of their regular revolution, theme is a transposition of substance, the character of the transposition being determined by the kind, quality, and peculiarity of the focal point; one effect being produced when two photos points come in conjunction; another effect by the conjunction of photos with scotos points.

The moon sphere is an aggregate shell of the reflex unity of earth and planetary spheres. It has its four focal centers or polar points on its circumference, as do the other spheres. Eclipses are dependent upon the relative positions of these poles to the poles of other heavenly bodies.

The light of an entire dynamosphere (sphere of substance), whether planetary or lunar, depends upon the relation of its focal points to the focal point of the helio-sphere--the sphere called the sun.

Every eclipse is produced by the crossing of the scotoic (dark) point, the pole or focus of one dynamosphere, across the photoic point or pole of another sphere.

The photoic and scotoic poles are at the opposite extremities of a given axis. If the photoic pole is toward you, the scotoic pole will be at the opposite extreme of the axes, and therefore opposite you; the photoic pole being on a line from you to the dark point. The movement of this dark pole or point across the light point of another sphere, cuts off the source of radiation to the sphere which suffers the eclipse.

Geometrical Basis of all Measurement; Limitation a Pronounced Property of Form

The geometrical basis of all measurement, namely, the unit, the duad, and the triad, must constitute the root of all mensuration and the central point of geometrical measurement and limitation. The limit of the shell of this sphere of universal life is the co-ordinate limit of material existence. Nor can geometry separate the sphere and the cube as the circular and cubical integralism of that whole comprising the material form of being, in which creation forever perpetuates itself within its own spheres of active performance.

When once it is known that there is a limit to the universal form of that material whole called the universe; that there is a definite center beyond which there is no material thing,--a center of space beyond which there is no space, it can then be easily understood that there must exist a co-ordinating and circumferential limit which of course must be the limitation of the sphere. The universe having a material basis, possessing all of the properties of all material things, must certainly embrace every property of form, because every material thing has some kind of form.

Limitation is one of the most pronounced properties of form; therefore, the material part of the universe, having the properties of all material things, must have limitation. The belief in the illimitability of the universe has its co-ordinate in the idea of a division of matter in which, though it has been conceived that there was an atom, and that it was indivisible, there is yet a possibility of the atoms being shivered into still other, lesser particles, and these into still greater minutia,--because the idea of divisibility does not meet the question of so called scientific definition as to illimitability.

There is a limit to magnitude; there is also a corresponding and co-ordinate limit to minuteness. There is a point of indivisibility (this is the geometric point, the minutest quantity of matter), and agitation at this point (the friction which it cannot resist) will reduce it beyond the material state to its metaphysical condition, in which there are none of the properties of matter; it is then the substance which, though no longer matter, is nevertheless as substantial as it was when in the condition or state of matter from which it was reduced, and from which it came.

We not only possess a knowledge of the correlation of the so called energies, but the correlation of the various kinds of material substance,-- that correlation which determines the interchangeability of two co-ordinate substances, matter and its coextensive and correlate ether or spirit, or its essence.

CENTRAL AND CIRCUMFERENTIAL LIMITATION. A UNIVERSE WITHIN A UNIVERSE

WE HAVE DISCLOSED the character of universal form. It is defined as having the two limitations--center and circumference. These are not only limitations as to space, but they are limits as to activities in space. Work or energy must be confined to these two limitations, which must be the extremities of all activities. These extremes must also be the extremes of the transformations of matter, and the points of the correlations of matter and its co-ordinate ether.

The flow of ether toward the center stops with the limitation centrally; the co-ordinate flow toward the circumference stops also with the limitation toward the circumference. The upward and downward ways, the universal anode and cathode of ethereal activity, come to their terminal extremities at these two extremes of center and circumference. The ethereal vibrations have many limitations between these two final extremes; hence the qualities and characteristics of ethereal activity are determined by the relation of these oscillations in the ethereal coruscations and graves.

There are degrees of attenuation beyond the matter condition of universal space, that is, in that domain where matter ceases to be, and which in the non-vital sphere may also be called metaphysical--beyond the physical. It is not, however, strictly in the realm of metaphysics, for it deals with the physical sphere of ether, which is distinctively differentiated from the thought domain, as in the psychic and pneumic realms of thought.

The greatest discovery of the age as to cosmogony is the discovery of the cellular form and character of the world in which we live, and which constitutes the confines of the material universe. In fact, it is the only great material discovery of this century. It settles the great question of origin and destiny, and all of the great problems that are agitating the mental world at this juncture of scientific and social revolution.

There is a universe within a universe, so to speak; by this, I mean an anthropostic existence within a purely physical form, in which there is a corresponding activity or function. Each is the reciprocal product of the other; both are interdependent, and together they constitute one. These two distinctive forms and qualities are the physical universe, which we have described, and the anthropostic universe of human and psychic existence, which is in and constitutes a part of the physical universe.

The physical activities which define the operations of both matter and the essence of matter, in the domain of physics, are quite different from activities which define the operations of anthropostic existence; but between the two there are correspondential and antithetical analogies which correlate the two domains.

For instance, in the physical universe there are stories in the heavens, one above the other in space, interior to which is the central star. In every story in the physical heavens there are astral nuclei belonging to each degree or story; but none of these is visible to the natural eye except those in the first or lower-most heaven.

The visible stars are in our own atmosphere--the lowest natural or physical heaven. These stars are not great worlds, but rather focal points of "energy," wherein there are partial materialization and dematerialization as the result of active combustion. Correspondentially, in the anthropostic heavens (the heavens within the human race, of which humanity is the rind or pediment) there are degrees, the first containing the mental centers, the most brilliant of which only are accredited with being stars. However, each mentality is a star of great or less magnitude,--or if not a star, a nebulous approach to one. As in the physical universe there is a central star, so in the anthropostic there is a corresponding stellar center.

Physical Heavens Constitute Pattern of Social Government

The order of the physical heavens, in which all of the stars are related to the central one and regulated by it, must constitute the pattern after which the social government is to be formulated. The order and regulation of the stars of every magnitude in the Physical heavens are determined by the relation which the central star sustains to all of the stars subject to its government. The government of the physical universe is imperial, in that the head of government resides in one center; but democratic, in that all of the stars bear that reciprocal relation which makes the center dependent upon the reciprocal activity of the subsidiary but contributary centers.

While there is a subordinate relation of the multiplicity of stars to the central one, so there is a subordination of the central star to all of the stars, whence the central one derives its powers of government. The regulation of society, therefore, is not left to another experiment, because former experiments have failed to accomplish for the people that for which government is established, but must be regulated by the scientific knowledge and application of principles which may be determined before the correct form of government is instituted.

The science of the Cellular Cosmogony, then, determines what the final form of social government shall be which, though not equal, will enforce an equitable relation of all of the gradations of social relation and activity. The center cannot heap up and enlarge, for if that were possible there would be engendered hypertrophy of the heart and center, by which all of the other parts of the organic structure would be thrown out of proportion and balance, thereby engendering disease that would be destructive to the entire organism. The central star would be in a state of active receptivity, but would be distributing to all of the parts the qualities and substances which it had prepared for redistribution.

What is true of the central star would also be true of every part of the entire economic system of the physical universe. Here, then, is found the scientific pattern of the forms and functions of the anthropostic world, which the physical world constantly gestates. As the center of the physical universe depends upon its circumference, so does the circumference depend upon the center. As this is true in the physical, so is it true as to the anthropostic.

LIFE DEVELOPS AND MATURES IN A SHELL, EGG, OR WOMB; HENCE WE ARE INSIDE OF IT

THERE IS EXTANT a theory that the earth is flat.

THE arguments employed do not necessarily prove the earth to be concave, but they do not prove the earth flat. The earth is not necessarily flat because it is not convex. There are ten thousand arguments at hand to demonstrate the truth of the concave or cellular theory; and every argument brought forward by the so called savant to prove the Copernican theory can be as easily demolished as the one considered.

We deem it important to simply announce the broad statement that all life develops in a shell, egg, or womb, and that the law of development in the greater or macrocosmic order does not depart from the universal law. All natural life develops and matures (to the point of its liberation from environment) in the egg or womb. The earth, therefore, is the great womb of natural development, hence we are living in a shell.

It is one of the modern miracles that the human mind, otherwise apparently so profound, can in its estimations ignore the law of foreshortening in the attempt to prove the convex rotundity of the earth, when it admits the law for all other purposes. It is as difficult to eradicate an error and impress a truth today as at any period of the world's history. Human progress advances upon the principle that "Where ignorance is bliss, 'tis folly to be wise."

In the illustration of the railway, it was observed that two rails four feet apart appeared to narrow down to the dimension of one rail. If the track or railway be cut in two at the point of vanishing, or at the point where the two rails appear as one, and all that part between the observer and the vanishing point be removed, so that the observer looks against the end of the two rails where they appear as one, there will be seen no intermediate space; and from mere observation without reflection it would be denied that two rails existed. Now let us imagine the observer to be a great scientific (?) teacher, and that he says to another:

"Do you see that rail yonder in the distance?"

"Yes," is the reply, "what of it?"

"That is a binary rail. I can make it look like two rails some distance apart."

"How can you do that? I don't believe it," says the incredulous neighbor.

"Why, just look here. I have an instrument that magnifies distant objects, and by applying its magnifying power I can separate that into a double object, and enable you to see two rails. This is proof that the rail is a binary one. These binary rails are the most curious of all rails."

Two balloons traversing space on parallel lines fifty feet apart will reach a point where they seem to blend as one, precisely as two rails separated by a space of four feet will seem to come together in the distance and appear as one.

"Do you see that balloon yonder?" says the scientific (?) investigator to his neighbor.

"Yes," answers the neighbor.

"That is a binary balloon," continues the scientist. "A binary balloon; what's that?"

"Why, a binary balloon is one which, when submitted to observation through a telescope, appears as two balloons. It's a phenomenal balloon. They are not so numerous as the single balloons, but may be seen under favorable circumstances."

On the basis of the supposition that space is illimitable, let us imagine two stars so far distant from the observer as to appear one, though a million (?) miles distant from each other. The million miles of space have seemed to vanish to the apparent contraction of the object which appears but a minute speck in the distance. It must be remembered that they are separated by a million miles of space.

If an elongated object could extend through that space, covering the million miles, its diameter as large as the diameter of the two bodies, it still could not be seen as more than the mere star point. But if it were a million and two miles at the vanishing point, it could be seen extending beyond, and enlarging the apparent point. The farther it receded in the perspective, the longer it would have to be made to be observed as a point.

Apparent Contraction of Space

Some years ago we were in conversation with an active and thinking mind, one familiar with the astronomical idea of the resolution of a star into binary and multiplex forms,--which the telescope is capable of effecting. In reply to our remark that at any distance beyond the vanishing point of a given space,--such as the four foot space between the tracks of a railroad,--objects must be outside the four foot limit, and that the further the distance was extended the farther apart they must be to be seen as if at the median line between the rails, he said:

"You do not pretend to say that two trees so far apart can be seen," marking the position on a diagram before us, "as one tree, do you?"

"This is precisely what we do say," we answered.

If the astronomers, instead of calling two stars (which they believe to be separated by millions of miles) a binary star, that is, a double star, would say that the appearance of a star is the result of the contraction of visual area, the apparent contraction of space so as to bring two stars to be observed as one astral center, thus ceasing to put the cart before the horse, this simple proposition would be understood when applied to terrestrial concerns, and much confusion of mind would be obviated.

If two stars (separated by a million miles) can be seen from a given point of observation as one star, it follows that beyond that point two stars of the same size, to be seen on the same line of observation, must be farther apart; and the farther distant they are the farther apart they must be to be observed on this given line, or, so to speak, given level.

"But," says the inquirer, "what is the Doctor driving at? What is he trying to prove?"

We reply, we are attempting to make the stupidity of this age awake to the fact that a pole or a mast must be elongated in inches proportionably to the square of the distance in miles, to maintain the top of a succession of poles or masts on an apparent level. We are trying to awaken the mind of so called civilization to the fact that, as an object recedes in the distance, it appears to contract at both ends by virtue of the law of foreshortening, and that that which is usually attributed to convexity of the earth is really due to diminution of visual area through perspective foreshortening. It's all in your eye!

Comets Are the Production of the Relations of the Sun's Motion to the Colures

The word comet is derived or Anglicized from the Greek and Latin cometas, and means hair. The comets are productions, of the relations of the sun's motion to the colures. The colures are the two prime meridians. The solar and lunar orbits are respectively related to these meridians. The term colure means docktail, or the tail cut, off. The points on the equator and at the tropics where the two prime meridians (the colures) cross, are the principal points on the ecliptic (cutting off) where cometic "force" is generated.

The comets are composed of cruosic "force," caused by condensation of substance through the dissipation of the caloric substance at the opening of the electro-magnetic circuits, which closes the conduits of solar and lunar "energy." This cut-off substance forms itself (according to circumstances) into lenticular shapes of various forms, such as double convex or convexo-convex, double concave or concavo-concave, piano-concave and piano-.convex, diverging meniscus, and converging meniscus.

These condensations of substance into lenses through which the sun's rays pass, sometimes cause refractions of light through them to appear as long trains, while it is nothing but the sun's diverted rays of light. They whirl through space in a spiral, approaching nearer the sun, until they enter the sun's vortex as one of the principal sources of solar supply.

The sun is a helix. Its motion through space--north and south, in that complex activity which occasions the seasons--is a spiral like that of an induction wire around the piece of steel in the induced magnet called the helix, from helios, the Greek for sun. While in its passage north and south the sun reaches its solstitial place at the tropics, its rays extend to the spherical limit and terminate around the poles in zones or rings of cruosic force, the motions of which are derived from the impetus of the sun's motion in its orbit at the solstices.

These rings of aggregated physical substance whirl around the poles at a rapid rate, and break at that point in either tropic where the sun enters and departs from his solstitial genuflections and bearings.

They then contract in their circular longitude, and attain the characteristic lenticular form which the relation of the break to the motion causes the rings to assume in their longitudinal contraction into lenses.

After breaking and contracting into lenticular form, they then start out in the spiral motion and orbit, ultimately falling into the sun, whence the substance was originally derived. At long intervals the same continuation of the sun's impetus and derived "energy" produces a corresponding ring, and another comet of the same order starts out in the same spiral, and is regarded by the astronomers as the return of the same comet.

The Lunar Function and Form

There can be no more interesting study relative to cosmogony and to luno-anthropology than that which is offered in lunar (Unction and form. It is the hylegiacal center which governs the principles of formulative creation. The lunar sphere is the great menstrual reservoir and channel of universal fluxion as pertaining both to alchemico-organic activity and the corresponding principles in the organo-vital sphere of creation. The moon is queen of the psychic realm, as the sun is king of the pneumic spheres. In this aspect of their qualities, the sex functions of moon and sun are viewed from the external or exoteric point of observation.

As the hylegiacal center and sphere of formulative force the moon holds, in the solutions of her menstrua, all the elements of transformation from which the foundations of the universe are laid and its super-structure established. She is the terminal of the seven planetary oozings, and the final reservoir of their accumulations. The basis and resource of her power to rebuild are the seven laminæ or beaten plates (rakayia) of the firmament, rendered stable through the processes of her depository function.

She is the final product of the action of solar substance upon the metallic strata contiguous and superimposed one upon another, comprising the outer rind of the crust of the earth, reflected as an energetic menstruum and aggregated as the lunar gravo-photosphere beneath the contiguity of the upper stratum of the oxygen of our atmosphere and the lower circumference of the atmosphere of hydrogen above us.

The subtle and interior forces of the sun penetrate the inner crust and water of the earth's surface, permeating even the metallic strata and acting as a disintegrator to the layers of metallic substance, reducing their surfaces of contiguity to electromagnetic and levic substances, which proceed, as a complex solution of menstruum, from seven metallic bases constituting so many circumferences, formulating in the heavens--as it proceeds from these circumferences toward the center of the cosmos--the seven planetary spheres.

The direct cause of the aggregation of the seven spheres or planets is the conjunction of the inflowing spirit-substances, of which there are seven qualities, with the co-ordinate seven qualities outflowing from the solar sphere. The moon is the culminating and aggregate product of the seven; she being the final receptacle of the seven fluxions.

The Moon's Phases

The waxing and waning of the moon are continuations of the same phenomena belonging to the planets. The moon is not a direct reflection of the earth's surface against the contiguity of our present oxygen with the hydrogen atmosphere above us, but the consecutive storage reflexions of the various planes of metallic strata responding to the penetration of solar "energy." We have in the moon a vague but correct outline of the surface of the earth, implanted by a storage process and viewed by us as a complex reflexion of the concavity of the earth. We see Europe, Asia, Africa, North and South America, Oceanica, the waters of the earth, etc., pictured for our inspection in outline above us.

Eclipses of the Sun and Moon

One of the principal proofs adduced of the globular form of the astronomi-cal bodies, is the fact that in an eclipse the supposed body passing between the one eclipsed and the sun forms a circular shadow. This would be positive proof if there could be adduced no other or better reason for the phenomenon. That is, if no other equally cogent reason could be assigned, this might be taken as proof; otherwise it is no proof.

The sun transmits its radiations in a circular form, as may be illustrated by the appearance of the rainbow. These radiations strike or touch the concave strata of the earth's circumference as only a circumradiation can do, and must therefore, in a reflex action of those emanations, return to the pivot or center of influx in a circumvergent, as they passed out in a circumdivergent form.

Ecliptical phenomena, whatsoever may conspire to effect them, must necessarily conform in contour, in the circumcision or cutting off, to the circumvergent aspect of the energetic fluxion, whether afferent or efferent in direction. If it can be determined by what processes the circuit is closed and the current generated, it can as readily be determined by what processes the circuit can be opened and the current eclipsed.

Purposes of the Ecliptic

Every phenomenon is governed by law operative for some specific purpose. We therefore study the laws of the ecliptic with the end in view for which they are instituted. The object of the ecliptic and the operation of its functions are the conservation of "energy" and the perpetuity of motion. "Except those days should be shortened, there should no flesh be saved," has direct reference to the application of the principles which govern the ecliptic in the alchemico-organic world, as well as those which govern circumcision and the direction of its uses as a religious rite; and the laws of conservation, operative in the alchemico-organic, are dependent upon those operative in the organo-vital, and are related to them as effect to cause.

Position of the Ecliptic (Cutting Off)

The ecliptic is the line or direction of the sun's yearly course. According to the Copernican system it is the earth's orbit around the sun, and therefore the sun's apparent annual motion. The earth is a shell, with its concave surface occupied. In other words, the surface we occupy is concave instead of convex, and is comparatively stationary. That which we call the sun is the projected focus of the occult or hidden solar center. His motion is helical or spiral from east to west, moving toward the south, in his gyrations, six months of the year, and north the other six months.

The limitations of these motions are the two tropics. The sun has no zenith point north of the tropic of Cancer, nor south of the tropic of Capricorn. His zenith at the tropic of Cancer is June 21, and at the tropic of Capricorn, December 21. These are called the solstices, meaning the standing still of the sun; for at these points the sun circles the earth without going farther north or south until making a complete diurnal circle. June 21, the rays of the sun are vertical at the tropic of Cancer; December 21, they are vertical at the tropic of Capricorn.

Influence of the Motion of the Sun upon the Metallic Laminæ and Surface of the Sphere

In the orbit of the sun there are four prime points or centers; photoic, scotoic, caloric, and cruosic. These four primary substances and influences follow one another in the gyre of the solar motion. Their action is as if there were four gyres successively following one another in the order of photoic substance (lumen, light), caloric substance, (thermos, heat), scotoic substance (the substance of darkness), and cruosic substance (crystalline or frigid substance) . Four distinct helices of "energy" are winding their course and exercising their co-ordinate and antithetical influences upon the surfaces they touch and the substances they penetrate, day after day, in the perpetual solar gyre.

Suppose we take the axis, the poles of which are heat and cold; the heat and cold points or poles being exactly opposite. The tendency of the gyre of calorine is to perpetually expand as Helios (the sun) winds his never-ceasing spire. Following this course, twelve hours behind, cruosine, the freezing substance, or the substance of crystallization, exerts its contracting force as Helios winds his way.

Here, then, we have the application of the law of pulsation, as regular as the expansion and contraction of the heart beat in the human body, and, from the corresponding law in the alchemico-organic domain, to that operative in the domain of the organo-vital sphere. We are not only enabled to observe the application of the principles of expansion and contraction, alternately applied in solar influence through the penetration of these solar substances, but we also find herein the law of insulation.

The radiation of heat is cut off in the direction of the cruosic gyre, reflected back upon itself, and compelled to take a lengthwise accelerated course, producing friction, and therefore the generation of magnetic substance of the terrestrial quality, as contradistinct to that of celestial origin.

FORCES AND FACTORS WHICH PERPETUATE UNIVERSE

THE revolution of the central star or stellar center which, from its positive and negative sides, produces the revolutions of the projected sun, precipitates also (in its revolutions) the essences of the sun, which come in contact with the inflowing substantial essences from the circumference. These unite in the atmospheres and appropriate their substances in processes of combustion.

In this union of outflowing essences from the sun and inflowing essences from the circumference, and the burning of the substances of the atmospheres, there is a constant precipitation of matter,--reduced from the state or quality of spirit to the state or quality of matter.

The earth's surface is thus constantly accumulating matter on its uppermost (innermost) surface, day by day, in the direction of the sun's apparent revolution around the earth,--which is really the projected sun's movement through the lower atmosphere,--in an orbit around the central star and within the crust or shell, the circumference of all.

The superficial earth and water--the water represented by the large body of oceanic mass--are conditions of intermediate metamorphosis or change from the condition of spirit and aerial stages of substance to the mineral and metallic states. We have referred especially to the seven laminæ and their polate centers.

The five mineral depositions or strata not directly represented by the known geologic formations are related to one another in mineral planes, and focalize their polate points in the five primary fixed stars, similarly to the focalization of the essences of the metallic planes in the centers called planets. The five earth planes, therefore, have five corresponding polations. Upon these seven polations of the metallic and five of the mineral spheres, depends the arrangement of all subsequent polations constituting the starry belt called the Zodiac.

The Zodiac is divided into twelve sections, supposed by modern scientists to be merely arbitrary divisions having no natural foundations. The peculiar mapping out of the heavens into constellations, and naming them according to the names of certain forms of animal life, are regarded as purely arbitrary and the result of the ignorance and superstition of the ancients.

We will here undertake to show that these tracings and mappings, or classification and nomenclature, are the result of the possession (by the ancients) of positive knowledges of the truth concerning not only the origin of the constellations, that is, of the focal centers, but of their special division into twelve segments, rather than the result of ignorance and superstition.

In order to make plain to the reader the laws by which this division is governed, we must insist upon an effort to at least constantly hold the mind to the conception of the intraspherical philosophy, or that which demonstrates that we live within the sphere or globe, as opposed to the current teaching that we live upon the convex or outside surface of the globe.

Origin of the Stars and Planets

The central star, the real polar point, which is the positive origin of the stars and planets, in transmitting its essences outwardly, forms around itself the circumambient space of light and darkness, or a positive and a negative side. Upon this peculiar arrangement depend night and day. The earth's surface is subject one half of the time, or about that, to the influence of the light side, and the other half of the time to the influence exerted upon the earth's surface by the darkness and its concomitants.

In this revolution of the central star, which gives to us the appearance of the revolution of the sun in an orbit around the earth once in twenty-four hours, and which, by the modern scientific "lights" is interpreted to mean the rotation of the globe upon its axis every twenty-four hours, there must be presented to the earth two directly opposite poles--the light and dark poles. Half way between these poles there exist two others; one is the evening and the other is the morning. These are the poles of twilight. The evening is the caloric or heat pole; the other is the cruosic or cold pole.

The revolution of the sun in one continuous direction causes the earth's encumberment of matter to follow as a consequence upon a perpetual spiral; not, however, a spiral of the same and persistent outfluence, but of a consecution of fluences modified by the specific effect of each polate point, as these points succeed one another in the order of the sun's rotation.

By this we mean that there are four orders of encumbering consecution; namely, midday, evening, midnight, and morning; and each of these points exerts its specific fluence upon not only the superficial surface of the earth, but upon its deeper surfaces also. The importance of this observation cannot be appreciated from a superficial consideration of the subject, especially when we take into account the earth or ground only, and the changes which take place there by the union of the ascending and the descending substances.

What the Vegetable Kingdom Affords

The vegetable kingdom affords one of the most favorable opportunities to study these subtle fluences,--the outfluences and influences characteristi- cally different at any two opposite polate points. In the foliage of vegetable life, the so called lungs of vegetation, there are carried on the double process and function, according to the period of the day to which it relates, of what partially agrees with the function of respiration in the lungs.

In the morning the leaf gives forth oxygen, and in the evening carbonic acid gas, or carbonic anhydride. These are the marked characteristic differences in the respiration of plant life as pertaining to the caloric (evening) and the cruosic (morning) poles. The specific characteristic differences in the respiration of plants at midday and midnight, while as thoroughly opposite and distinct as the differences in the evening and morning, are of a more subtle character because partaking more of the nature of the transposition of spirit than of the more tangible substances--the gases.

The foliage of the vegetable kingdom, it will be seen, performs more than the single function of respiration corresponding to the respiration of the lungs. In the respiratory function of the lungs there is a constant union of oxygen and nitrogen inhaled with the carbon, which constitutes the base of the venous corpuscle, and really constitutes the fuel for the process of combustion that is in constant operation. The carbon carried into the lungs by the venous circulation enters into an actual process of combustion, uniting with the oxygen which is inhaled by the respiratory action of the lungs.

This union is not merely an absorption of oxygen, and therefore an oxygenation of the venous corpuscle, converting it to an arterial corpuscle, but it is the union or marriage of the white and the venous corpuscle by which is developed the red blood or arterial corpuscle. The carbonic anhydride exhaled or breathed out is the one product of combustion. This, as one of the offices of the lungs in the process of respiration, corresponds to the process which takes place in the leaf at night.

What the Capillary System Affords

At the extremity of the arterial circulation there is a process the reverse of that which takes place in the lungs. There is a process of combustion in operation in the capillary circulation which, instead of transmitting outwardly the spirit corresponding to carbonic anhydride, which is exhaled by the lungs, carries it back into the venous circulation, thus carbonizing the blood and supplying it with its sugar, the foundation of the carbon corpuscle.

In the animal structure the process of oxidation takes place at one extremity of the circulation, namely the lungs, and the process of carbonization at the other extremity. In the vegetable kingdom the processes of carbonization and oxidation take place at the same extremity; namely, in the foliage, but at the two extremes of the day.

By this critical observation we see that morning and evening are the two extremities of a cycle or revolution, and that the foliage is related to every degree of this revolution, and represents the entire cycle. The revolution of the day has its four polate centers, and of course its intermediate segments of the cycle. The leaf represents this cycle complete.

In the animal kingdom, of which man (who constitutes the microcosm) is the representative, we have noticed two extremes; that which corresponds to midday, (viewing the lungs from their office as performed toward the circulation, and not as to exhalation,) and midnight, represented by the other extremity of the circulation. We thus define the polate centers in the microcosm, corresponding to the two supreme points of the revolution.

Relation of Microcosm to Macrocosm

In the relation of the microcosm to the macrocosm there is developed an important discovery; namely, that the motion is inversely to the motism and statism of the physical macrocosm. For instance, in the physical macrocosm the vegetative and vegetable form and function are stable, and the diurnal relations are mobile. In the microcosm, the vegetative form and function are mobile, and the diurnal relations are stable. It is thus discovered that the stable things in the microcosm are the mobile things in the macrocosm, and that the mobile things in the microcosm are the stable things in the macrocosm.

We have presented the vegetable kingdom as an illustration of the action of the four polate points or centers, having defined especially the two prominent poles and their fluences upon the function of respiration as exhibited in the plant. If the plant exhales and inhales, the zone or sphere of revolution (as related to the plant) has a complementary inhalation and exhalation inverse to that of the plant. This exhalation and inhalation must be specifically and correspondentially active at the four poles described, to correspond to, complement, and co-ordinate the activities of the vegetable respiration.

Vegetation alone, while exhibiting the phenomena in a marked degree, is not the only department of the physical circumference subject to and modified by these subtle fluences. The metallic and mineral deposits, the various earths, rocks, salts, etc., also the water over the surface of the earth, constantly inhale and exhale to meet specifically the fluences of these polate centers. The substances transposed in the form of gases and in the conditions of various essences are invariably the consequences of the combustion in operation in the earth, water, and air.

An Unsettled Problem

It is claimed that the surface of the earth moves from west to east nearly twenty-five thousand miles in twenty-four hours. According to modern "science," this is a solid body moving through space at a rotary speed of over one thousand miles an hour, or more than sixteen miles per minute. Can any sane man imagine that a solid body with this rate of speed, surrounded by a thin atmosphere, can so carry its atmosphere with its momentum as not to produce a contrary motion of the atmosphere, while at the same time it would cause the rotation of an oscillating pendulum?

No one disputes the fact of the motion of the pendulum, as first observed by Foucault, and later experimented with by Flammarion. We might, however, question the uniform direction of the oscillating pendulum in a series of experiments. But allowing the experiment to be fair and the motion as reported, we would inquire, What causes such a phenomenon?

Whatsoever the cause of the motion it must be considered with regard to two propositions;--first, the supposition that the earth revolves because the heavens appear to revolve; and the motion of the pendulum is taken as corroborative testimony to an hypothesis, a guess, which still hangs in doubt with the astronomers, for the reason that with howsoever much reinforcement you sustain a guess, it still remains hypothetical. And the astronomers will not stop seeking for still further corroboration, because they are still in doubt. We wish to assure our readers that the problem is not settled under the Copernican system.

The second proposition is that the Koreshan Geodetic Survey has settled forever the fact that the earth is a concave shell, and that man inhabits the cellular sphere. If it could be proven that the earth rotates, then the pendulum would act the same on the inner as it would on the outer surface of a ball, were it the motion of the earth that caused it. It could not, therefore, affect the fact of the Cellular Cosmogony in the least.

The Earth Comparatively Stationary

According to the Koreshan System, the earth is comparatively stationary and the heavens are moving within the stationary earth. The sun is moving at the rate of about eighteen thousand miles in twenty-four hours. It is sweeping through space with this velocity and radiating its "energies" into the environing shell, in which there is a corresponding magnetic spiral motion.

To this spiral motion of "energy" is due the rotation of the pendulum, and not to the motion of the earth. First, it will be understood that the pendulum is suspended from a support attached solidly to the body of the earth. Second, it will be noticed that the curve of the earth is practically the same at both extremes of the oscillation, the earth moving just as rapidly at one point as at the other. There could not be a calculable commensuration of difference either in time or space, at the two extremities, as to curvation or the time of longitudinal motion.

If the pendulum is swung from north to south and south to north at the start, it would be subject to the eastward motion of the earth, which, were the theory of the earth's rotate impression in relation to the pendulum ball true, the ball would apparently move toward the west, and with the opposite swing of the pendulum it would swing equally toward the west-- the motion on one side balancing the other. This would be the effect if the phenomenon were the result of the earth's motion.

How would it be when the pendulum rotated around to the east and west points? The earth would be rotating toward the east, the pendulum swinging east and west. The earth is moving, according to the Copernican hypothesis, at the rate of sixteen miles a minute either with or against the ball, while it swings westward, and at the same rate with it when the ball swings eastward. If the earth by its rotation affects the motion of the pendulum enough to cause this rotation, why would it not make an appreciable difference distinctively marked while swinging east and west?

The swinging of a pendulum could bear no possible relation to the earth's rotation, even if the earth were a ball rotating from east to west at the rate of twenty-five thousand miles in twenty-four hours. . . . The marvelous thing about this experiment, is that any man possessing any claim whatsoever to the title of scientific, should accept this solution without asking the question, "May there not be some hypothesis for this motion, as reasonable as, or more so than, the hypothesis of the rotation of the earth?"

If a pendulum were swung at the north oscillating laterally across the plane of the earth's rotation, (were there such a motion of the earth,) the pendulum hung in space (not upon supports solidly in the earth), there would be some sense to the proposition; as it is it is the veriest nonsense, and later the "scientists" will laugh at their own folly.

THE ONLY TRUE SCIENCE IS FOUNDED ON A DEMON-STRATED PREMISE

THE Copernican system of astronomy (built upon an hypothesis) seems to satisfy the minds which will not and cannot thing. The Koreshan Astronomy has its foundation on the rock. We present an experimental fact, the result of invention and months of careful labor by the Geodetic Staff of the Koreshan Unity, thus furnishing unquestionable proof of the Cellular Cosmogony advocated by the discoverer since 1870.

In view of the fact that the so called scientists declare "That hypothesis or guesswork, indeed, lies at the foundation of all scientific knowledge," we maintain the right to call a halt, while we declare from positive experiment . . . that the only true science is founded upon a demonstrated premise, and not upon an assumption or mere guesswork. We know we are right. We know, as well, that no man can guess at a premise and claim that he knows the conclusion reasonably; for it is admitted by all scientists that their premises are hypothetical.

It has been said that the man who cannot think is a fool; the man who will not think is a bigot; and the man who dares not think is a coward. It matters not to us what excuses are offered for ignoring us, . . . we can show to the people--by methods so simple that all can understand--that we are right, and that the astronomical guessers are wrong, and, through the common masses, compel the humbug scientists to admit their folly and blundering hypothetical processes.

Schiaparelli discovered seas, and also land covered with canals on Mars. More critical (?) observation may discover that the canals on Mars traverse the seas, and the scientist (?) changes the conclusion of yesterday to an opposite conclusion today. Yesterday he observed seas on Mars; but today what he declared to be water is land, because the Martian contractors have run the canals across what the other scientists (?) declared to be seas or oceans. Now we ask in good faith, Is this science (knowledge) or is it guesswork? Are these hypothetical conjurers, scientists or quacks? And we ask, How long must the world be gulled by the sham and nonsense of pretended experts in science?

We are absolutely sure of our ground; and we reiterate, that no man who builds a theory upon the basis of an hypothesis, and who declares that "hypothesis or guesswork, indeed, lies at the foundation of all scientific knowledge," has any reasonable claims to consideration. Guesswork is not science.

A statement made by a so called scientist today and contradicted tomorrow is not science; and yet this is the stuff that has been palmed off as scientific upon a credulous public for hundreds of years, and the men who have the audacity to eject these emanations are dubbed as scientists.

"I cannot see why the universe should be limited to a single cosmic cell. The analogy of cell structure in the human body, with groups co-ordinated in interdependent series, would seem to suggest a plurality of worlds, limited in number, and forming, in their serial aggregate, that larger cosmic structure which Swedenborg designated as the 'Grand Man.' Does the logic of your premise inexorably limit the universe to a single world?"

We are frequently met with the above inquiry. The cosmic cell, which we claim includes the universe, focalizes its universal imprint upon myriads of stellar points through the subtension of its pencilings of potency. These focal stellar nuclei are grouped in clusters according to the geometric action of reflection and refraction. These constellations, fixed in their positions and relations according to the "inexorable" laws of geometrization, correspond to nations and individuals of humanity; and in the lesser form of creation (the microcosm), to the arrangement of cells in the infolded or incubated form of the vidual.

Were the ordinary human form opened out or evolved into the form of the cosmic shell, as it is before incubation, it would be in the form of the grand cosmic structure, with its rind or circumference and its stellar groups, though in magnitude the correspondent of the microcosm or the little universe,--the universe in its least form.

Every star in the grand cell (the universe) is the imprint of the whole in proportion to its attitude toward all other stars, and is complete in proportion to its approximation to the astral nucleus. The stellar nucleus is the point at the center of the great camera obscura, where the photograph (light writing) is taken of the great shell and all contained in it.

As the astral center is the photograph of the alchemico-organic (physical) cosmos as an entirety, it is (in the least form) the kinetic nucleus of the essences of the cosmic structure, and therefore the point of both the inception and exception; that is, the point of the limitation of the afferent tendency, and beginning of the efferent flow, that is, of radiation.

Astral Nucleus of the Physical Cosmos an Eternally Fixed Point

This astral point is not the Lord God, but it is the point in the alchemico-organic whole which co-ordinates with and corresponds to the astral nucleus in humanity; that is, to the Lord. The astral nucleus is an eternally fixed point near the center of the alchemico-organic structure, because it is in and related to space.

The corresponding stellar nucleus in humanity appears and disappears as the Lord, the Son of God, at stated periods of the world's progress. These changes of state with man correspond to the varied qualities of stellar nuclei in the space of the alchemico-organic cosmos.

When man attains to the perfection reached by the Lord Jesus, he is so related to all things in the natural and spiritual humanity as to render him as central to this whole as the astral (alchemico-organic) nucleus is central to the alchemico-organic cosmos. For this reason, when the crucifixion of the Lord obstructed the flow of the anthropostic nucleus, the current of the physical cosmos was interrupted and the sun was darkened.

A principal lesson is found here in the general law of astrology. The central man (the Lord Jesus Christ, the bright and Morning Star of the anthropostic cosmos) was so related to the central star of the alchemico-organic world as to interrupt its currents (its vibrations) when the current of his humanity was for the time being obscured.

Descending or Gravic, and Ascending or Levic Eliminations

In the Koreshan Cosmogony it is announced that immediately above our common atmosphere of oxygen and nitrogen there obtains one of pure hydrogen. In the activities which comprise the life and perpetuity of that atmospheric field there are combinations of spirit and matter which, if not precipitated, would constitute deleterious elements. These are thrown down and at once appropriated by the field below.

The gravic (descending) essences produced by the contiguity of the hydrogen with the oxygen upon which it rests, unite in our own atmosphere with co-ordinate levic essences produced at the point of contact of our atmosphere with the surface upon which it rests.

The descending or gravic eliminations of the hydrogen sphere combine, in our atmosphere, with the levic eliminations of the surface below; and in the union of the two the cloud is formed and thence the water is produced, which, precipitated (dejected as water), clarifies the atmosphere and becomes a supporter of life in the field beneath. There is not a domain in existence, either in the alchemico-organic world or in the lower biologic, anthropostic, angelic, or theo-anthropostic, in which this law does not prevail and operate.

The hells themselves generate the forms of life they cannot endure, and spew them upward in their ascending flight to realms of joy above. Christ the Lord in his ascending development came from the hells, and was the first begotten of the dead; and when created, had he remained a tangible personality among the inhabitants of earth, would have done so as an effete element of the nether world, a cause of disintegration, a disease in the body politic and a disrupter of society; but being eliminated as an ejection from the sphere of brutality whence he arose, he operated as the conservator of the less brutal field of benign activity in his spiritual power.

Cosmogonical Limitations as to Time and Space

The physical universe is limited by virtue of the fact that an organic whole could not exist without all of the properties of form. The physical universe has form because it is a material structure governed by operations which have their co-ordinating influences at the central pole of regulation. One of the properties of form is limitation. Without limitation there could be no form of the universe.

Everything in the universe flows toward and into the center; and inversely, from the center into the circumference. These relative fluxions are mutual and compensative, and in their reciprocal interfluxions they maintain the equilibrium of the circulations of the universe, which makes it an eternal, self-recreative structure.

Time limitations are governed by definite cycles, which are determined by the solar, lunar, stellar, and planetary motions. All cycles are primarily determined by the momentum of the stellar nucleus, from which all other motions are derived. The momentum of all the operations and activities of the universe is imparted from the reactive forces of the stellar center, which derives its momentum from a continuous influx from the circumference, which is a complex and composite material form with a corresponding function.

The shell of this universe (which is sole and unique as the entire universal field of activity, and which constitutes being) is about one hundred miles thick, composed of seven outside metallic strata, superimposed upon which is an inner shell, rind, or crust of mineral layers, of which there are five; and upon this are the geologic strata. These seventeen layers or strata comprise the rind of the material universe, of which, in its involuntary and evolutionary processes, the perfect man is the highest power of development.

The universe has an anatomical structure like the anatomy of man, with the exception that with man the incubated form is manifest for the convenience of human activity and use. The incubated form of the universe is the form of man. This is the macro-cosmic man. The microcosm is the man in the least form of the universe.

Relation of Central Star to Circumferences

The motions of the universe, by which times are determined and perpetuated, depend upon the motions of the central star as related to the circumferences which reflect the forces which the central nucleus radiates. The rotation of the central star--which, because of its positive and negative reactions is forced into an excentric revolution, and thence into a spiral--determines first, day and night, and thence the seasons.

The seasons are produced by the spiral motion, which moves the astral nucleus or central star north and then south about forty-seven degrees; that is, twenty-three and a half degrees north and south of the equator. The cause of the motion is the reciprocal action of the inflowing essences of dematerialized matter, with a co-ordinating outflowing radiation reacting and from the rotary motion thus induced, merging into the spiral.

The electro-magnetic center is made to heap up within its helical cone of electro-magnetic essence, a surplus "energy" (rather electro-magnetic substance) until the motion toward the north, or toward the apex of the cone, is discharged into the solar zone. The star then starts backward toward its base of the conic helix, or toward the southern aspect of its motion. The helical motion thus engendered constitutes the eternal spiral momentum of the universe.

The precession of the equinoxes (which is about fifty seconds of a degree every year) determines one cycle, which is about twenty-four thousand years, because there are processes of the foreshortening of time which reduce what would otherwise be twenty-five thousand eight hundred and sixteen years, to the limit of twenty-four thousand years. This period of time carries the sign through all of the twelve constellations of the Zodiac, returning the sign to its own house (or to the constellation Aries) at the end of the twenty-four-thousand year period.

There is a co-ordinate movement of the ecliptic on the equator, which embraces a period of seventy-two thousand years; and one of the ecliptic on the solstitial colure, which embraces twelve distinct periods of twelve thousand years each, including in all a period of 144,000 thousand years.

These time periods on the solstitial colure are determined by catastrophal times, caused by a sudden movement of the ecliptic thirty degrees on the solstitial cycle. The world is now preparing for one of these movements. The one now impending will carry the ecliptic down to seven degrees below what is now the equator. The earth will be shortened in its longitudinal axis proportionably, and the sun will then become a zone encircling the earth, making light all of the time, modified by one half of the annulus being less bright than the other half. Then there will be no night, as Scripturally predicted.

There are twelve sudden movements on the colure (solstitial), one in twelve thousand years. These are all accompanied by universal catastrophes. One of these movements as noted above is almost due, and we are about entering upon one of the world's greatest phenomenal periods.

PLUMBLINE IS FIRST STEP IN RATIONAL DEMON-STRATION

The Sun Is Constructed on the Basis of a Helix

AFTER "scientific" men reach the position in mental concept and conviction that the earth is hollow, and that earthquakes are the result of the vibrations of a shell composed of layers of metals--placed one upon another in contiguous succession--beaten out by the processes of Nature's pulsations to form the rind or pediment of the superimposed atmospheres, and that volcanoes are the result of chemical pustules sometimes produced by the igneous union of natural gas, petroleum, and coal mines, in this rind or skin of the great hollow sphere, they may take one more step and apply the true laws of analogical construction and discover that we are on the inner surface of the sphere.

The world is hollow. In the physical sense every principle of reason confirms and demonstrates this belief. This is not all. The surface occupied by man is concave; and though so called scientific men m butt their heads against the adamantine wall of truth, succeeding generations will look back to the GUIDING STAR and FLAMING SWORD as the harbingers of true wisdom regarding cosmogonical construction.

By slow processes and roundabout methods the world gradually gropes its way through the darkness toward the dawn of the coning day; gradually the light reveals the forms of life, and enables them to be studied in the light and application of genuine science. The plumbline is the only true first step in every rational demonstration. This is the first scientific element in the hand of the Koreshan, in the formulation of the trigonometrical calculus of demonstration.

"When I wake in the morning and cast my eyes toward the east, I see a tree, forty rods away, and the rising sun, about six thousand miles distant, at the same instant. If it be true that a substance has to leave my brain and extend to a distant object, by means of which an impulse is carried back to the brain, it travels with a speed beyond my comprehension, and just as quickly to a distant object as to a near one."--[Excerpt from letter to Koresh.]

Koreshan Cosmogony holds that the central sun is less than four thousand miles away and invisible to us, and that the projected sun is at the point of the conjunction of our atmosphere with the atmosphere of hydrogen resting upon it. So that objection is obviated. Light or visual force is rapid, but it is not impossible for the mind to distinguish the difference between its communication at short or long distances.

I have stood upon the shore of Lake Michigan and in the distance, say about three miles from Chicago, have observed a permanent object. I closed my eyes for a few seconds, with their direction toward it. Instantly upon opening them I observed near objects, but it required about three seconds for the distant one to come into view. Visual "energy" is a thousand-fold more rapid than electrical "energy."

The sun at the center, or comprising a solar limbus to the astral nucleus, is constructed upon the basis of a helix. Like the astral center around which it forms a hemisphere, it has a light and a dark side. Its axis is inclined to the circumference of the earth, or to the earth's axis, in the same proportion as in the commonly accepted theory of astronomy, the earth's axis is supposed to be inclined to her own orbit around the sun.

The sun having a light and a dark side, is the recipient of an influx of spirit from the dark circumference, or semi-circumference, while it is projecting from its light side the "energy" of light to be focalized, in its projection through atmospheres and spheres of "energy," at two points; making in all three distinct suns; one for the highest atmosphere, one for the middle atmosphere, and one for the third, last, and outermost atmosphere. The last one focalized is the one which shines in our own circumference, and is visible to the natural eye.

As the astral center revolves upon its axis, its projections from its photoic side (being focal points from this side) must necessarily move in orbits around the astral nucleus. The peculiar relation of the astral axis to the earth's circumference, and therefore to the orbits of the projected suns, causes them to move in spirals, north and south, determining the seasons. As this motion is as if there were a process of winding, like the winding of thread upon a bobbin or spool, or like the wire upon a piece of steel, as in the magnetic battery, the Greeks called the sun helios; to wind in spirals.

The Sun a Great Magneto-electric Battery

The sun in its relation to the earth is nothing more nor less than a great compound magneto-electric battery, generating distinctively six primary "energies;" namely, light and heat, one pair; electricity and magnetism, the second pair, and levity and gravity, the third pair. These are respectively, in the order named above, photoic, caloric, electric, magnetic, levic, and gravic "energy." They are merely what were material substances reduced to the most subtle solutions. Though they are "energies" and substantial,-- composed of what had been atoms of matter,--they are no longer material, but spiritual. If an atom of matter is destroyed as an atom of matter, it at once becomes the spirit of that quality or kind of matter, and while just as substantial as before, is no longer material.

The astral center, with its concomitant solar system, revolves upon its axis; the earth being relatively and comparatively stationary. At the center of the system, this being about four thousand miles from the circumference or concave habitable surface, is a peculiar formation resulting from the emplacements of "energy," disposed or arranged by co-operative activities of refraction and reflection. This arrangement assumes the form of a tabernacle and tent, more nearly described by the Scriptural exposition than can be expressed in any other form of language.

Such cosmogonical construction is in harmony with all forms of creation, and has the advantage of being in agreement with the laws of development as everywhere observed, wheresoever the order of growth comes within the scope of observation and reason. In this system we have the great cell or egg of development, the progress of growth corresponding to the general law of incubation.

We are not begging the so called conservative people of the world to even examine the Koreshan System of Cosmogony. We do not fall upon our knees to fogyism. We have the true theory of construction, and know whereof we affirm. We declare our doctrine, knowing it will gain adherents from the thinking and reasoning people who are looking for some positive and tangible expression of the truths of Deity.

Moon the Compound Reflection of all the Strata. Liquid Mercury the Intermetallic Substance

It will be remembered that the moon is reflected from the strata comprising the metallic crust of the sphere. The action of the sun upon the earth is in reality the action of the sun upon the moon. The moon is not the reflection of any single stratum, but the compound reflection of all the strata. The penetration of the thermal and cruosic rays into the strata, causing the alternate expansion and contraction of the metallic laminæ, observes a spiral course in the laminæ, corresponding to the gyral motion of the sun.

As the heat expands the metallic substances the spaces between them contract; and as the cruosic substance contracts the laminæ, the spaces between expand. The result is an outward spiral current of the substance which fills the interstices between the laminæ.

The menstruum filling the vacuities, and which is being pushed along through a continual spiral from north to south and from south to north, between the tropics, or over forty-seven degrees of the earth's laminæ, is mercury (quicksilver), holding in liquid solution the elements of the intermetallic channel.

The motion of the sun is not merely a spiral north and south, but a spiral, enlarging and diminishing itself alternately, having a maximum and minimum field, or circuit of motion; hence there are alternate periods of approximation to, and remoteness from, the concave surface of the earth.

This approach of the orbit to, and departure from, the earth is the phenomenon called by astronomers perihelion and aphelion; from peri, around or near; and apo, distant or away from, and helios, the sun.

In Koreshan nomenclature it would be called the sun's perigee, near the earth; and the sun's apogee, distant from the earth, as indicating the nearest and remotest points of his approach and departure as he describes his helical orbit.

The cause of all motion resides, primarily, in the voluntary principle of the perfect human (God) mind. We say the God mind, referring the reader to the mind of the God-Man, the illustrious Christ of God, in whom was the fulness of the Godhead bodily; God in him having attained the ultimates of his being, he constituting the esse and existere of Deity.

Voluntary action begets the involuntary, its antithetical co-ordinate. The supreme cause of motion is in desire; and the supreme desire is love toward God as a function of the ascending man, and the love of God toward man, as the function of the descending attraction of God. These two co-ordinate attractions result in conjunctive unity of the two, and God and man become one.

This law of motion is all pervasive, being let down by gradation through all the degrees of motion until its potencies operate outwardly into the alchemico-organic world. It is therefore seen that all the motions of the alchemico-organic, while originating in voluntary thought, are not the direct and immediate operation of mind upon those domains of activity; but there is a correspondence between the two, and the analogy is so perfect that a correct interpretation of the alchemico-organic will furnish, through correspondence, the correct interpretation of the anthropostic.

Cause of Perigee, Apogee, and Helical Motions

The proximate cause of the perigee and apogee of the orbit of the sun resides in the laws of expansion and contraction, induced by the alternation of heat and cold as follows: Heat is the result of friction; where there is the more resistance there is the more friction; and where there is the more friction there is more intense combustion. There can be no exception to this law.

The thermal substance of the sun is most intense at the vertical point of radiation, less intense as the rays are more oblique, and least intense at the lateral ray. This would be true even though the heat were measured at points of equal distance on every line of divergence. The pole opposite the vertical ray would be the coldest point.

Let us suppose the central and vertical substance of the sun to be potassium. The direct action of this ray would not constitute a thermal ray; but if this spirit meets, in its radiation to the circumference, the converging or afferent flow of cruosine, or cruosic substance, the resistance produces the friction from which proceeds the heat, precisely as flame will proceed from the union of potassium and ice.

Just as we have the north pole and the equator where two opposite conditions obtain, so we have the north side of the sum and the south side, where opposite conditions also obtain; and the alternation of these attitudes alternates the sides of expansion and contraction. This relative action produces the deviation of orbital motion.

The actinism of the sun's substance as he is caused to approach to or recede from the concavity within which he revolves, is successively specific upon the metallic laminæ which his substances penetrate, subjecting them to the successive alternation of heat and cold, applied to the contiguous layers, penetrating first the strata nearest the surface of the earth, and successively reaching the more outer layers until he acts upon the outermost.

Outermost Metallic Plate of Earth's Crust the Greatest in Specific Gravity

That it may not appear (in this solution of the moon problem) that the discussion of the operations of the sun comprises the more prominent factor, we will here reiterate the statement that the moon is the product of the influence of the sun's activities upon the terrestrial strata. We cannot, therefore, discuss the origin, form, and function of the moon independently of a general and specific consideration of solar functions and phenomena.

The reader is already familiar with the fact that the crust, shell, or rind of the earth is composed of contiguous laminæ or strata, concave in form, in seven primary metallic plates, superimposed one upon another; that which is greatest in specific gravity constituting the outermost plate, while the others are arranged according to diminution in the ratio of their specific gravities.

The operation of the sun's gyre (spiral motion) in the penetration of his essences into these metallic crusts acts specifically upon them, primarily, according to the quality of the physical spirit, whether it be photoic, scotoic, cruosic or thermic;--these being his primary substances.

The penetration of the thermal physical spirit must assume the form of a circular impression upon the laminæ, and must move in a spiral or gyre in the direction of the gyre of the sun as he winds his helix north and south.

The phenomenon following this action of the thermal radiation, manifest in the laminæ, would be singular in this: That as heat expands more where most intense, and less where least intense, the plates would become thickest at the vertical penetration (that is, where the thermal ray was perpendicular to the central radius), and thinnest at the circumference of the radiation.

Hence, between two plates (laminæ) pressed together by the process of expansion the interstice would be filled. If, twelve hours later, there follows this process of expansion and closer contiguity of the laminæ, a process of contraction by virtue of the action of cruosic physical spirit, a circular concavity would follow the spiral course of the obliterated interstice. This concavity being filled with mercury, there would necessarily move a circular disk of mercurial solution in a spiral course from tropic to tropic.

This would provide an amalgamated surface for each of the laminæ, acting at once as a conservator of the superfice and intrafice of the contiguous laminæ, and as an insulator and channel for the magnetic current generated in the activities of the solution and the lamina.

The Sun Has Secondary Gyre

Added to the common and primary gyral or helical motion of the sun in his annual course north and south between the tropics, he has an axillary motion around an axis perpendicular to the concavity of the earth, hence the solar substances are disseminated in a spiral, and this momentum is imparted to the mercurial discus, which, in addition to its motion with the solar helix, revolves from the impetus of the imparted solar axillation.

At Gordon's Pass.
(Group of Witnesses and Position of Apparatus at Farthest Point South.

There are four primary laws of motion originating and moving as follows: The first impulse from combustion is radiatory; this meets the counter and resistant moment [motion] forming the circular, which, in a second resistance, is transformed to the spiral. The impact of the radiatory with the resistant, convergent, or afferent flow of physical spirit produces the undulatory or coruscatory movement.

With the secondary solar gyre, as with the primary helix, there are four primary polar points corresponding to the caloric, cruosic, photoic, and scotoic nuclei; and corresponding substances are radiated toward the metallic circumferences. From these centers there are secondary disci of mercurial solution formed in the inter-metallic spaces, which, by the secondary solar gyre, are caused to move in orbits around the primary discus in some of the planes.

Between the outer laminæ, upon the gold stratum, instead of there being formed a number of disci surrounding the primary discus, the vermiculation (peristaltic motion) is less complete, and the disci merge into rings of mercurial solution. The secondary disci of some of the inter-metallic laminæ are reflected into the heavens as so called moons of the planets (the "moons of Jupiter" are from such sources), and the mercurial rings as rings of Saturn.

The radiation of the solar substances toward and into the laminæ is not direct from the Solar center to each of the circumferential strata. The radiation of physical spirit from one stratum to another, through all the seven laminæ, is successive, observing a graduated scale of transmission; the ratio of increase being a geometrical formula mathematically governed by the complex square of the ratios of specific gravity and places of deposition.

The operation of these laws, comprising the principles of both motion and form, would impart the peristaltium to the strata (laminæ), which continues in them after the direct action of the solar radiation has passed over the plates.

There is a primary mercurial discus between each pair of strata. Each discus pursues its spiral course, moved by the thermal physical spirit along the track mapped out by the course of the solar gyre. When we consider the fact that the disci are moved along their spiral course upon the surfaces of these seven metallic laminæ, and associate this fact with the fact that the momentum diminishes (from the inner to the outer discus) with the square of the complex ratios above noted, we are supplied with the data from which may be accounted that specific relative motion of the planets, wherein those of the inner orbits overtake those of the outer, and the laws of their annual circuits propounded and elaborated.

The planets proper are general aggregations of physical spirit heaped up through the reflection of the solar physical spirit from the metallic laminæ. The substances from these aggregations converge to the astral nucleus and are thence planted, through this nucleus or focal point, by a succession of divergences and refractions, upon the mercurial disci, and are again reflected from these and impressed upon the planetary stratifications in the heavens.

The Peculiar Office of Mercury as a General Solvent

Thus far, we have considered only the specific action of the two antithetical substances (caloric and cruosic) upon the laminæ and disci, as effecting the peristaltic progress in the gyre of their circuits. The observation of these depends upon the action of the photoine and scotoine, or the light and the dark substances moving in their gyrations, respectively, between the calorine and the cruosine impulses.

The photoic physical spirit acts specifically different from either calorine or cruosine. It has a subtle power of penetrability into and through the mercurial disci, imparting to the atoms comprising the compound solution held in amalgamation in mercury as the basis of the solvency, differential motion, as each quality reacts against the penetrating photoine.

That the above may be clearly comprehended, it will be well to recall to mind the fact that mercury attracts to itself (as it passes along, washing the metallic surfaces) the metallic atoms loosened by the action of the thermal and other substances, and absorbs and dissolves them.

The mercurial solution is consequently a general solvent for the metallic substances through which it passes; therefore, as the photoine penetrates the discus it imparts a precipitate motion to the general substance in solution, for the atom of each kind receives a motion of its own in resistance to the photoic impulse.

The passage of the mercurial solution is not confined to any single interspace; for at the tropics and the equator (where the ecliptic and equatorial circles meet) there are openings for the evacuation of the menstrua from the various interspaces, and their discharge into other interspaces.

While the menstruum of one cavity is making its passage through one of the metallic interspaces, it both attracts to itself the substances of the surfaces to which it is exposed, and makes certain depositions to the surface through which it is passing, of the elements derived from its passage through a former one.

The continuous spiral canal through which the menstruum is impulsed by the action of the solar substances in the alternate expansion and contraction of the metallic strata is, so to speak, a sort of alimentary canal, and corresponds, in the alchemico-organic cosmos, to the alimentary canal of the human body; the functions being correspondentially the same. There is a correspondence also in the number and form of the divisions.

Photo-Alchemic Action Determines Color Resistance

That property of actinism through which the photoic reagency is manifest is largely influential in the determinations of the metamorphosis or transmutation of metallic elements; but it is not the only factor of the mutative processes. Every pigmentation is the result of the reagency of scotoic and photoic substances, and while it adds greatly to ornate attractiveness, this is not its only function. Processes of assimilation are dependent upon coloring as well as upon other factors of assimilation.

The character of the motion imparted to an atom of matter by photo-alchemic action upon the particle is determined by the resistance of the color, (each color offering its specific resistance,) being differently agitated, hence more or less rapidly metamorphosed and, therefore, differently posited.

No two atoms of matter of a given kind, going to make up the bulk of a mass, are differently posited in the mass without having yielded to different qualities of the same kind of force entering as a factor into the disposition of the atoms. Variations of shade in coloring, so slight that they could not be detected by the eye, would be sufficient to determine different depositions of the atoms.

That the reader may not labor under any false impression regarding the transmission of the solar substances, it will be well to state here that what we have denominated a thermal ray becomes the essence of heat only when a descending physical spirit of one kind meets the essence of an opposite kind. The heat is generated at the point and time of meeting.

No two substances can meet and produce their effect except as they form their conjunction and correlation in the form of matter adapted to the union and transmutation to be effected. Let us take, for illustration, the process of the formation of chloride of sodium in the ocean.

How Chloride of Sodium Is Formed

Sodium in minute quantities is constantly conveyed to the waters of the ocean, or any inland sea having no outlet, and transformed to chloride of sodium through the descent of solar essence. While it may not properly be called chlorine essence, the descending substance (meeting a co-ordinate ascending essence) does, in its union in the atom of sodium, produce chloride of sodium; and because there is no outlet to the ocean or sea, the solution accumulates.

We have entered but briefly into the exposition of the principle of photoine (light), in its action of differentiation in the process of deposition and assimilation. The mere presentation and study of the subject for a knowledge of the fact, would not be worth the while of the student and investigator. It is only when we apprehend the bearing of such knowledge upon life itself as pertaining to our relationship to God and to one another, In the fulfilment of uses to the neighbor in the performance of which we insure, by reflex action, the greatest use to self, that the joy of acquisition is experienced.

Precisely as light differentiates, selects, and rejects, with darkness as the background of resistance and impression, so does truth differentiate between good and evil, with fallacy as the background to insure contrast and enable the truth to direct in the acceptance of good and the rejection of evil.

The scotoic pole follows the photoic in the progress of the sun's gyre, and brings its influence to bear upon the particles differentiated and directed by the operation of light, not in a direct manner, but indirectly through its influence to obstruct or hold in rest the substance that did not directly respond to the influence of photoine.

As the determination of photoine accelerates differently each quality of atom upon which it reacts, when scotoine reacts it must perform its function to retard the various atoms where it finds them, this being at different places, because the momentum of photoic acceleration has given to each quality a different impulse from every other kind.

In the reagencies of photoine and scotoine, we possess the properties which co-ordinately determine the emplacement of substances in the order of strata; therefore, the law of stratification. The compactness of the strata is determined by their compression through the alternate action of expansion and contraction, which is a process of beating the metallic substances forming the strata into thin, hard plates denominated, in the description of creation given in Genesis, rakiya; in English, rendered firmament.

We have already compared the motion and current of the laminas and disci with the alimentary canal in the human body. This not merely because there is a similarity in their motions, but because the alchemico-organic universe is the correspondent of the Grand Man, or the general anthropostic world; and in the doubling up of the contents of the cellular cosmos, in the process of the incubation of the great egg or cell of the universe, that part of the environment of the egg becomes the alimentary channel of the dispensations, and these correspond to the alimentary canal of the vidual man.

Candid Investigation Would Soon Convince a Truth Seeker

The mind that conceived the Copernican system, arising in the dark ages, was so simple as to take an appearance for a fact, and to deduce a theory which better thinkers of more modern times would very soon dispose of were it not for the fact that they will not take time to think.

We have shown by repeated illustrations that the convex theory cannot be true; but the astronomers and scientists, with their inconceivable bigotry, having gone crazy through the hallucinations of mediæval times, must lose their reputations as scientific men (and their bread and butter also) if they permit the world to be set right with regard to questions upon which they have built up for themselves names, and, through this, are honored as great lights and educators of the people.

If any man with brains, having facts at his command, will give two hours' candid and unbiased thought to the investigation of this subject, with the application of the principles of foreshortening as set forth in the literature of the Koreshan Unity, he will be convinced of the truth of the Cellular Cosmogony and of the utter absurdity of the so called Copernican hallucination.

We are told that we do not exhibit the spirit and principle of the Christ, regarding the so called scientists, when we resort to language sometimes seen in THE FLAMING SWORD. What difference does it make whether we say, as did the Lord, "fools and blind," blind fools, blind idiots, or blind bigots and idiots. He told the truth, because it was the best way to exasperate people and set them thinking.

Koreshanity has the truth, but the present humanity (steeped in tobacco, rum, and sensualism) prefers to meet the truth of Koreshanity with ridicule rather than to give it candid consideration. But as ridicule is always the strongest argument, and the one that usually has the greatest weight with the non-thinking mind, it is not surprising that the so called scientist usually takes refuge behind it when meeting a rational force that otherwise is irresistible.

TELESCOPES MORE OR LESS IMPERFECT AT BEST; LIKEWISE THE LENS OF EVERY HUMAN EYE

NO ASTRONOMER pretends to believe that a telescope can be made which does not embrace more or less of the uncertainty of astigmatism. No optician lives who does not know that the lens of every human eye has more or less the uncertainty of astigmatism; how, then, can an observation be made, involving the immense distances pretended to obtain by the men who believe in the Copernican system of astronomy? If correct calculations are made of relative locations of stars in view of these uncertainties, what comprise the factors of certainty?

This may be answered, first, by the statement that the stars are within our atmosphere instead of beyond, therefore the factor of refraction does not enter into the problem; second, the base line is concave instead of convex; and the lines upon which the relative location of the star is made, in the determination of parallaxes, are convergent toward the object which is observed within the atmosphere, and at a short distance from the point of observation.

We were asked the question: "If the experimental operations of the geodesists should not prove the Cellular Cosmogony, would that invalidate your claims regarding other parts of your System?" We replied: "Our system of theology, sociology, and government is founded upon the fact of the concavity of the earth. We are as positive regarding the truth of the inside theory now, as we shall be after all the corroborations of the Geodetic work." The importance of the Geodetic operations resides in the fact that we shall have in hand the results of practical mechanics applied to the earth's contour, so that no man can refute or gainsay them.

A direct line extended at right angles from a perpendicular post strikes the earth. This we have demonstrated absolutely. That the earth is concave, there is not the shadow of a doubt. It destroys the last ray of hope entertained by the scientists who, upon the basis of the Copernican system, either deny God, or what is worse, say they do not know whether there be a God or not.

In extending the air line we have proven, by simple mechanical application, the concave theory. We have found the ratio of curvation to be about eight inches to the mile. This is the first time in the history of the world (so far as known) that a like or corresponding measurement has been taken for the determination of the contour of the earth. We are enabled to assert positively the undeniable fact of the Cellular Cosmogony, for which we have contended many years.

We place positive demonstration against assumption. We know what we are talking about. We know that the so called scientists know that what they are presenting to the world is merely pretense, for not a scientist living pretends to claim that the premise of the Copernican system is anything more than conjecture, an unproven hypothesis.

The First Step of the Copernican System Is Mere Hypothesis

We have suspended the plumbline. From this we have extended, at right angles, an air line which strikes the earth proportionately to the height of the perpendicular. This experiment can be repeated a thousand times with the same results. There is not a loop-hole for the poor deluded advocates and devotees of the heathen system of astronomy which holds the world (the Christian world) in the chaos of midnight darkness.

Astronomers know that the first step of the Copernican system is a mere hypothesis. They know that it has not been demonstrated, and that any system resting upon an assumption, or a piece of guess-work, is liable to fall to the ground. Our first step is not an hypothesis. We know our ground. We not only assert that a straight or air line extended at right angles from a perpendicular post will touch the earth in any direction it may run, but we know it because we have made the experiment, and have found the results precisely as we declared they would be.

We expect to find greater obstacles in the way of public recognition than did Columbus in his efforts to carry forward his project of discovery, for greater issues hang in the balance of this adjustment of human belief. It is a death-blow to the modern Christian church, for if the church which pretends to be imbued with the Holy Ghost, or the Spirit of Truth, can permit the world to be so entirely ignorant of the first principles and laws of creation as it now is, it is a holy spirit hardly worth cultivating.

THE GEODETIC SURVEY

Unreasonable Opposition to Our Claims

IN THE STUBBORN RESISTANCE immediately manifested by a few who would have rejoiced had the evidences produced by the employment of the Rectilineator been favorable to the Copernican theory we find history repeating itself. Because of this we publish the following quotation from the pen of a recognized scientist, which reveals the character of the opposition to the facts of demonstration and observation in its true light. Fallacy is rooted and grounded in the very heart of humanity, and will not abandon its hold without a struggle.

I learned, as my first great lesson in the inquiry into obscure fields of knowledge, never to accept the disbelief of great men or their accusations of imposture or of imbecility, as of any weight when opposed to the repeated observation of facts by other men admittedly sane and honest. The whole history of science shows us that whenever educated and scientific men of any age have denied the facts of other investigators on a priori grounds of absurdity, the deniers have always been wrong.--Prof. A. Russel Wallace, the Eminent Naturalist.

For the benefit of the reader we here recall the idiotic opposition to the facts. of the discovery, by the noted Harvey, of the circulation of the blood in the human system. Although he produced the evidence,--made ocular demonstrations of the flow of the blood through the arteries and veins of the human body, the physicians and anatomists of the old schools refused to investigate. Today, there is not a man that can be found to deny the circulation of the life fluid through the arterial and venous systems.

When the telescope was invented and began to reveal the movement of satellites about the planets, the facts observed by Galileo and others were stubbornly denied by the astronomers of the Ptolemaic system; and for years Galileo succeeded in inducing but few to witness the phenomena through the telescope.

One scientist who had more zeal, prejudice, and jealousy than knowledge and wisdom, wrote a dissertation on the telescope, attempting to show how astigmatisms could be produced in the lenses, and the lenses made to revolve in such a way as to give the appearance of the satellites revolving around the planet Jupiter. That man lived and died without having made a single astronomical observation by means of the telescope, What did he know about the telescope? The sequel proves that he knew nothing; yet he denied that the objects that were seen, were possible to be seen.

The earth's concavity is considered as an absurdity, and the long line of demonstrations of the same, the mere result of deception and fraud. What do our critics know about the facts we have observed? Upon what reasonable ground can the evidences we present be disputed by those who have never undertaken the lines of experimentation we have projected? The opposition to our work today is as unreasonable, absurd, and idiotic as that manifested against the work of Harvey and Galileo.

We persistently proclaim the facts we have observed and obtained; they are as persistently denied by a few who have never ventured near the surface of water to test its contour. We have surveyed a line by means of a mechanical apparatus, the results of which are as easily (?) explained away by a few who have never seen the apparatus, and who know nothing about its capabilities or methods of use, as the moons of Jupiter were explained by those who had never observed their motions through the telescope. These so called explanations of the results of our survey we purpose to overthrow.

We have surveyed a line upon the Gulf coast of Florida. The measurements were such as to demonstrate conclusively the concave arc of the earth's curvature upon which the survey was made, and only the direction of its curvature, but also its ratio. In this survey we found a definite ratio of approach of the earth's surface to meet the rectiline extended from an altitude of 10 feet above the water level. The line was extended into the water's surface at a distance of about four miles from the beginning of the survey, as shown in the diagram accompanying this paragraph.

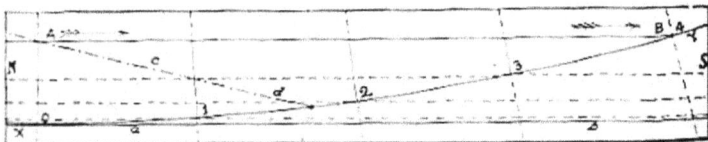

XY represent the concave arc over 4 miles in length; AB, the air line, A, the beginning, ten feet above the water, and B, the point of extension into the water; N is north, S, south; 1, 2, 3, 4, the mile stations; ab represent the external tangent parallel to the air line, showing the ratio of curvation of the earth's surface for each mile of the survey, while cd represents a line referred to later in the article.

The ratio of the concave curvature was in proportion to the square of the distance; at the end of the first mile the distance between the air line and the water's surface was eight inches less than at beginning, because the earth in this distance had curvated upward eight inches; the second mile, about two feet and eight inches; third mile, six feet, while at the end of the fourth mile the line extended into the Gulf.

In the diagram, the lines of perpendiculars extending from the air line to the arc at 0, 1, 2, 3, 4, decrease in length in precisely the same proportion that the earth curvates concavely. Not only was the proper ratio found to exist at the end of four miles, but also at the end of every eighth of a mile from the beginning of the survey.

Let each reader capable of making a mathematical calculation of the ratio of curvature of a concave sphere 25,000 miles in circumference, test this ratio and the results obtained by the survey, and it will be found that such ratio cannot be obtained by the extension of a right line upon any other than a concave surface; the geometrical principles involved will not admit of it.

Extra Precautions Were Taken to Overcome all Objections

We confront two classes of objections to the character of the Geodetic work upon the coast of the Gulf of Mexico. From several sources it is claimed that the Rectilineator was not sufficiently accurate to extend a straight line; while from others comes the accusation that we purposely inclined the first section at the starting point so as to extend the line into the water at a distance of four miles; from others, that the first section of the apparatus was not accurately leveled, but inclined toward the earth by mistake. We suppose these two classes of objections seem satisfactory to the minds expressing them.

We knew the objection would be urged that the apparatus was not accurate, and therefore took extra precautions, not only that such objections might be overthrown, but also to insure the accuracy required for such work; we did not devote four weeks to making the apparatus accurate for nothing.

The method employed to insure further accuracy was by making the apparatus neutralize its own inaccuracies by reversal or turning-over of each section at every alternate adjustment.

This process would correct any error arising from any inaccuracy of the brass-facings--for whatever error in the line would result from a single cross-arm being more or less than .005 of an inch out of right angle, would be corrected when that section should be reversed, as every mechanic well knows.

A source of inaccuracy is also attributed to the contraction and expansion of the material of which the apparatus is constructed. Those who make this objection have never seen the apparatus, and consequently cannot appreciate the fact that the plan of its construction obviates the effect of whatever contraction or expansion occurred.

Besides, there are no sources of error or inaccuracy--those of adjustment, settling, vibration from the wind, or change of temperature, that could conspire to produce a deviation of the air line always in same direction; check up all the errors that occur, as is done by all surveyors, and the value of the "elements of chance" is found to reside in the fact that the deviations are finally neutralized.

It is supposed that settling played an important part in the descent of the line surveyed; if so, why should the line descend .15 of an inch for the first eighth of a mile, and 6 inches for the eighth between the 19th and 20th tide stakes? If settling produced the descent, this would be manifest by returning over the same line. We returned over the same line for a distance of ⅜ of a mile, to ascertain if there would be any deviation.

The fact that the horizontal axis of the apparatus projected the line on the return survey to the same points on the record stakes indicating the air line in the forward survey, is proof of the fact that the factors of settling, if any existed, were absolutely neutralized, for they did not swerve the apparatus from a true and directed rectiline course. Let those who make such objections explain how the exact and definite ratio was obtained, if we did not extend a rectiline from the beginning of the survey.

Objections Contradict Each Other

We now come to the examination of the charge that we purposely inclined the first section so as to permit of extension of the line into the water in four miles. Our burden at the present time is not that we failed to produce the evidences that the earth is con-cave, but to get such minds to see the utter absurdity of such objections. The fact that these objections contradict each other, is conclusive proof that both classes of objections are made without a foundation of conclusion, and are simply subterfuges with which to evade the evidences afforded through the accurate survey.

Suppose that we did purposely incline the first section out of level, what would be the result? The charge involves the admission of three things,-- very necessary factors in the work of extending a rectline. First, that our mathematics was exact--necessarily so, to calculate the angle of inclination; second, that we were capable of making some absolute measurements of angles in the adjustment of the first section of the apparatus; and third, that the apparatus, in order to extend a line from the inclined position of the first section into the water at a distance agreed (?) upon before the work began, would have to be perfect and capable of extending a straight line, for with what else than a perfect apparatus and accurate measure-ments of angles could we strike the water at the distance desired from a given inclination of the first section from an absolute level?

In order for the charge to be true, we would have to extend an absolutely straight line, involving just the kind of adjustments and minute measure-ments that the other class of minds say is impossible. We know that the first section was level, having applied two of the finest levels obtainable.

We made no mistake--the accuracy of our line depended upon getting the first section in the absolutely correct position.

Reply to Charge of Inclining First Section

We will examine the charge in another light--in the light of geometrical principles, and will endeavor to illustrate our exposition of this charge and objection in the two diagrams in the following cut: Let XY represent the convex arc; AB the air line, beginning at an altitude 10 feet [1] above the water, and inclined out of level so as to strike the water in four miles; N is north, S, south; 1, 2, 3, 4 indicate mile stations along the line of survey.

Now contrast the ratio of approach of the line to the water's surface that we have exhibited in the cut illustrating the line extended over the concave arc. In the first diagram, the line began parallel with the water's surface, and ended obliquely to it. In this, the line begins at a definite angle from the horizontal, and ends coincidental with or parallel to the horizontal at B.

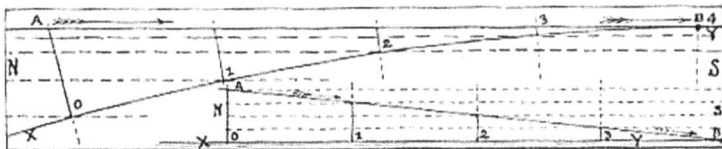

The ratio of approach of the line to the water's surface would be just the reverse to that really obtained. As for instance, beginning ten feet above the water, at the end of the first mile the line, according to the basis of the charge, would be four feet nearer the water's surface than at the beginning.

The ratio of approach of the line to the water's surface would be more rapid at the beginning of the line than at the ending, and the ratio of divergence of the line from the end of the line to the beginning would be in

---

[1] Really, to conform to the fact that the air line extended Into the water in four miles at an angle, with the horizon 1½ miles beyond the end of the chord, the altitude of the starting point, on the convex earth as per charge, would have to be 18 feet and 8 inches, instead of 10 feet! For the sake of simple illustration, we have represented the line as terminating at the horizon, and consequently parallel to the water.

proportion to the square of the distance from the end of the line; whereas in fact, in the survey of -the line at Naples, Fla., the ratio of convergence of the air line and the water's surface was less at the beginning of the line and greater at the ending.

If the air line had really descended at the ratio of four feet for the first mile, and so on in proportion, the line would have extended into the surface at a distance of 1¾ miles from the beginning of the line, and the angle of inclination from the horizontal would have been enormous.

If the earth were a plane, the result of inclining the first section of the Rectilineator so as to extend the line into the water in four miles would be an even ratio of descent of the line, as represented in the diagram below the convex arc in the above cut, in accordance with the principle of convergence of two straight lines.

XY would represent the water line; AB the air line as per charge; N is north, S is south; 1, 2, 3, 4, the mile stations. The approach of the line to the water for the first mile would have been 2½ feet; for the second mile, 2½ feet; and the same for the third and the fourth. The air line would sustain the same obliquity to the horizontal throughout the line of survey.

On the convex arc, the deviation of the horizontal axis of the apparatus would be greatest at the beginning of the line, and horizontal at the end. How was it? The line showed no perceptible deviation from the horizontal until near the end of the first eighth of a mile, and increased continually as the line progressed, until at the end of 2⅜ miles the obliquity of the air line to the horizontal was sufficient to extend the visual line from that point to the surface of the Gulf, below the horizon.

The above replies to the objections will be conclusive to those who will consider and comprehend the facts. It would be impossible to satisfy a man who doubted the accuracy of the multiplication table, until he came to a knowledge of some simple principles of mathematics. We do not expect to reach those who will not or cannot reason; their minds will have to be changed by the turn in the tide of popular favor from the old to the new system.

But it seems strange to us that--in the face of the facts obtained; in view of the fact that no such direct means have ever been applied by the old-school scientists; in view of the fact that they admit that all the so called evidences of the earth's convexity are only cumulative and circumstantial; and in view of the fact that those who criticise our work were not brave enough to face the issue and participate in the experiments performed by the Geodetic Staff--the direct and absolute evidences we have produced should be rejected by any mind capable of realizing the geometrical principles involved in simple right angles.

Corroborative of the demonstration of the earth's concavity by means of accurate survey, we have the long line of evidences obtained by other means,--tests of the surface of water on canals, lakes, and seas--tests and experiments which can be repeated upon any body of water to the satisfaction of the skeptical.

The facts we have observed and the line we have surveyed demonstrating the earth's true form, are susceptible of test. We challenge contradiction; but our challenge will have to be answered upon the field of contest, to which we dare the scientific critics, in the many lines of experimentation through which we have obtained the facts we announce to the world.

TESTS THAT CANNOT BE IGNORED

Practical Experiments Made by the Koreshan Geodetic Staff

AS THE FIRST of a series of proposed practical experiments with the view to demonstrating the true form of the water's surface, experiments were conducted upon the surface of the Old Illinois Drainage Canal, July 25, 1896; beginning with the bend in the canal at Summit, Ill., and running up the canal 5 miles to the northeast, to the first bridge.

Old Illinois and Michigan Drainage Canal. Site of Experiments, July 25, 1896.

At the beginning of the line of experiments, a target 22 inches in diameter was fixed upon a staff driven in the bottom of the canal, so that the center of the disc was just 18 inches, and its lower edge 7 inches, above the water. From this point a boat containing the three observers, with telescope, materials for sketches, etc., was rowed a distance of three miles; the boat was then anchored and an accurate view was obtained of the target.

The whole of the disc of the target was plainly visible, appearing a little above the water, with all the sections of colors of black, white, and red painted upon it. According to the accredited convexity, with the telescope 12 inches above the water, only 5 inches of the top of the disc should have been visible.

When the boat was rowed to the distance of 5 miles from the target it was anchored under the bridge, and another view was obtained with the telescope 12 inches above the water; the target was visible, also the hull or body of a barge located by the side of the target, upon which, at this distance, men were seen working.

At this point also, 5 measured miles from the target, the telescope was lowered to within 6 inches of the water, and through it the target and the barge were as plainly visible as with the instrument 12 inches from the water's surface; the target being plainly discernible against the bank of the canal, in the beginning of the bend in the course of the canal.

With the instrument six inches above the water, the horizon or apex of the bulge, on the basis of assumed convexity, would be about three fourths of a mile away, from which apex the water would curvate away for the remainder of the 5 miles; only three fourths of a mile of the water's surface could be visible to the eye unaided, or aided with the telescope.

The declination in the remaining 4½ miles would be 12 feet; the top of the target, which was 29 inches above the water, should have been 9 feet 7 inches below the line of vision; consequently, not only should the target be entirely invisible, but also the bank of the canal below the tow-path, which was less than 8 feet above the water.

Observations on Return Equally Convincing

Under the bridge from which the above observations were made, two large targets (one 21x27, the other 26x38 inches) were fastened side by side so that the lower edge of each was 7 inches above the water. The paper of which they were made was white, and they were placed in the sunshine directly beneath the bridge.

When the boat was rowed three miles on the return trip, observation was made with the telescope 12 inches above the water; the entire surface of the targets was plainly visible above the water.

Upon returning to the first target, 5 measured miles from the bridge, the boat was anchored; the sun was shining brightly upon the paper targets under the bridge; the targets were visible at this distance to the unaided eye of each observer in the boat, the eye being about 30 inches above the water.

The canal was quiet and still, with scarcely a ripple on its surface; the conditions were the best and most favorable for the final tests and observations of these experiments. A particular observation was made without the telescope. As the eye came within 15 inches of the surface, the targets became invisible; upon raising the eye again they came into view. Repeatedly the eye was lowered, but each time the targets could not be seen.

To the unaided eye, about three feet of space above the water appeared occulted, and that much of the piers under the bridge appeared out of sight. Will the telescope bring the targets into view again at a nearer approach to the water? Had a boat been alongside the targets it could not have been seen with the eye alone; the body of a barge three feet above the water would have been invisible.

The telescope was placed 12 inches above the water, and through it the targets were plainly visible. The instrument was then lowered to within 6 inches of the surface; the same view was obtained, with the entire surface of the targets in plain view.

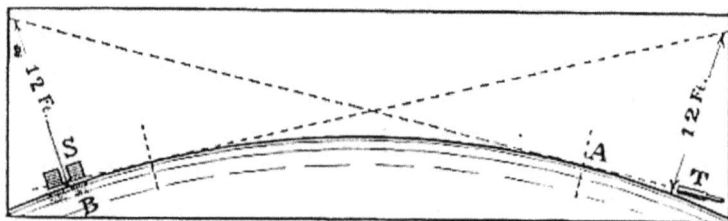

The result of the comparison of the conditions of observation with the accredited convexity is the same as in the case of the first target from the view under the bridge, with the instrument 6 inches above the water. The tops of the targets, if the water were convex, should have been 9 feet 10 inches below a direct line extending from the eye over the apex of the bulge to the terminus of the five miles. The accompanying diagram illustrates what would be the conditions and relations of the eye, the line of sight to the occulted objects, upon the basis of the calculated convexity. T represents the telescope, 6 inches from the surface; A, the apex, ¾ of a mile distant; S, the signals or targets, and B their reflections upon the water beneath.

Unmistakable Evidence of Concavity

But the most striking feature was noticed in the last observation at the end of the return journey; important, because it afforded the most unmistakable evidence of the, water's non-convexity. Directly beneath the targets were seen their white reflections upon the water, elongated and waving with the slightly rippling surface. We found here a fact mirrored in the water, which cannot possibly be explained away.

This view, obtained from careful and steady adjustment of the telescope, showed conclusively that not only were the targets seen, but also the water directly beneath the targets. Every foot of the water's surface between the anchored boat and the white targets was visible, also the surface of the water extending up the canal to bridge No. 2,--1½ miles more distant. The timbers to which the targets were fastened and the stones of which the piers were built were visible down to the surface of the canal.

The evidences presented in these observations were most satisfying and convincing. Manifestly, had there been the slightest convexity upon the surface of this canal, with the telescope 6 inches above the water, no reflexions of the targets upon the water beneath could be seen; and with the accredited convexity, any object under the bridge, 12 feet above the water's surface, would have been invisible.

The bridge, the piers, the bank on either side, with the two lines of telegraph poles, and the targets upon the water were carefully observed, as to their relative size. The last view through the instrument, of the relations of size and dimension of the objects, was the same as in the view with the unaided eye, one half mile from the bridge. There was no distortion; there could have been no refraction nor mirage. For comparison, carefully-drawn sketches were taken of each observation.

The same views can be had under similar conditions, with the targets and objects standing out in bold relief and in plain view, as indisputable testimony to the truth of the Koreshan Cosmogony, and in refutation of the modern system of science built upon the assumption of the water's convexity.

Once more the telescope is used to revolutionize science! Three hundred years ago, it was with the greatest difficulty that scientists could be induced to look through the magic tube; at that time, observation through it meant conversion to the new system. Today, this scientific instrument is put to a new use, and the principal difficulty now is to induce leaders of modern scientific thought to use it upon the surface of any body of water.

EXPERIMENTS ON LAKE MICHIGAN

IF THE WATER were convex, when boats and ships disappear in the distance hull down they would do so because the intervening hill of water would prevent their being seen; it is conclusive that if this were the case, the telescope would be powerless to render the occulted portion of the ships visible again. It is equally clear that if the telescope can restore the vanished hulls, the water upon which they sail is not bulged, and does not curvate downward beyond the horizon.

That the surface of the sea is convex, is shown by the way in which the ship disappears when it sails from the shore. First, the hull goes down behind the horizon, then the sails, and finally the mastheads. If the ship moved on any other than a convex surface it would appear again in the telescope.-- Prof. Peabody, Astronomer.

It is with reference to this phenomenon that special observations were made on August 16, 1896, from the shore of Lake Michigan, World's Fair grounds, by the Experimenting Staff. The atmosphere was clear and the horizon sharply defined against the sky beyond. Several sloop yachts and a schooner were observed at a distance of about 12 miles. From an altitude of 10 feet above the water (from a pier extending into the Lake), the hulls and about one half of the height of the masts were visible to the unaided eye. Through an opera glass, all of the surface of the sails and the full height of masts were visible, with the hulls still invisible; but with a telescope of about 50 powers, the hull of each vessel was brought into view with remarkable clearness.

We then went to the beach, and with the unaided eye about 30 inches above the surface of the water, only a very small portion of the top of the masts could be seen--they appeared like mere white specks just above the horizon.

With the eye this distance from the water, if the water were convex, the horizon would be two miles away, leaving 10 miles to curvate downward from the horizon, placing the hull of each boat 60 feet below the horizon.

As the masts of the sloop yachts were probably not over 40 feet in height, their tops would have been at least 20 feet out of sight.

It was now that the test came with the opera glass and the telescope. With the opera glass, only about one half of the height of the sails and masts could be seen; but through the telescope, the hull of each yacht, at the distance of 12 miles, was made plainly visible.

On the Lake shore at Roby, Ind., August 23, 1896, we made seven specific observations, some of which we briefly present below: We were greeted with the most beautiful horizon--clear and well defined; and the observations were rendered the more satisfactory by reason of the sunshine upon the vessels from the west.

As we approached the shore we observed, with the unaided eye, what appeared to be a mere white speck upon the horizon. It was a small steamer, with only a small portion of the pilot house visible above the water line. In the field of the telescope, applied to this horizon point, we observed the steamer down to the actual surface of the water upon which it rested; the whole of its body was in plain view.

In about half an hour the top of the smokestack of another smaller steamer was seen; and through the telescope the whole of the body of the vessel. A number of observations were made of some yachts, whose topmasts only were visible above the water line by means of the naked eye, but whose hulls were clearly seen through the instrument, the altitude of which was about 18 inches above the Lake level.

Presently, we saw a larger vessel running along the horizon line, with nearly the whole of its body out of sight. It was one of the liners running from the docks of Chicago to Michigan City, Ind. With the telescope it was brought into full view. It was going in a direction that soon took it entirely out of sight to the naked eye; not even the smallest portion of it was visible to the eye alone; its direction could only be pointed out by the cloud of smoke which followed it. Once the smoke cleared away, and there was nothing to indicate to the unaided eye the whereabouts of the vessel; and it could only be found by sweeping the horizon with the telescope.

To obtain the very best observation possible, the telescope was adjusted very carefully and allowed to rest upon a support; and through the steady atmosphere upon the quiet Lake we observed the whole of the vessel, every part of which was entirely obscured to the unaided vision.

The steamer was at least 15 miles distant; according to the accredited convexity, the lower part of the vessel would have been 150 feet below the horizon. If we considered refraction to be one third (it is seldom allowed to be over one fifth), there would remain 100 feet between the refracted visual line and the hull of the steamer!

Now let the facts of this observation be considered. If it be admitted that convexity intervened between the eye and the vessel to cut it off from view, would not the convexity still remain to occult it in the telescopic field? It is clearly to be seen that if convexity were the cause of the disappearance of the vessel, it would be as impossible to see it through the instrument as with the naked eye.

It seems strange that a matter so easily observed as this should have so long escaped even the most casual observer,--to say nothing of the scientist. We offer at this time no explanation of the reason they have overlooked it; suffice it now to say that what we have observed can be seen any clear and calm day upon the Lake.

Observations with Opera Glass and Telescope from Naples Beach, Florida

The following are some of the many observations made during the five months' experimentations at the Operating Station.

On January 12, from the Naples beach, a four-mast schooner was observed at a distance of about 10 miles. It at first appeared to be only a dark line upon the water horizon. Soon, however, it became more distinct. The hull and about one half the masts, and consequently about two thirds of the mainsails, appeared to be cut off by the water beneath. Through the large mounted telescope, not only were the top of the masts and the sails, but also the hull in plain view.

Capt. Gilbert, of the sloop Ada, who was present at the time of the observation, considered the vessel about half mast "down." Upon viewing the vessel through the instrument and seeing the hull even down to the water upon which it sailed, he considered it a genuine case of bringing into view the hull of a ship invisible to the unaided eye.

How do we know that we saw only about half the masts with the naked eye? We will endeavor to illustrate the same by means of a rough draft of the four-mast schooner in the accompanying cut. The topsails, it will be noticed, incline from topmasts to width of mainsails, leaving Vs between the topsails. The appearance of the vessel to the unaided vision was as in Fig. 1, the space between the topsails reaching almost to the horizon. The horizontal dotted line AB in Fig. 2, shows to the naked eye the apparent relation of the topsails to the horizon.

Fig. 2 shows the vessel as it appeared in the telescopic field throughout the time of the observation. This gives the comparative observations with the eye and telescope. An opera glass of about six diameters was also used, its power being sufficient to bring nearly all of the mainsails into view, but not the hull.

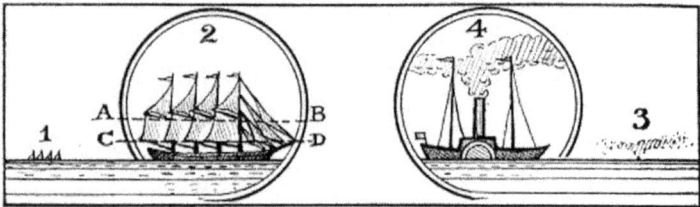

Through the opera glass, that part of the vessel above the dotted line CD could be seen. The more powerful the means of vision the farther the horizon is extended.

On January 19, about 4 p. m., we observed a dark cloud line just above the southern horizon. As a small funnel seemed to connect the cloud with the horizon, the conclusion was reached that it was a steamer coming into our horizon. There was nothing visible to the unaided vision above the water line but the smoke.

The appearance of the horizon at that point we have endeavored to represent in Fig. 3, in above cut. When viewed through the telescope, the lighthouse tender "Mangrove" was observed. The body of the vessel to the water line, the rigging, masts, and pilot-house were visible as shown in Fig. 4.

Away Down on the Standards
At Gordon's pass, 2½ miles south of Naples; 4 feet nearer the water than at the
beginning; the Straight-edges and the Horizon.

This observation was in every way satisfactory, because the condition of the atmosphere admitted of a sharp, clear view of the horizon and the smoke, which made the contrast more effective in the comparative observations. In this experiment the telescope brought all of the body of the "Mangrove" into view, when entirely invisible to the naked eye.

February 7, 5 p. m.; two-mast schooner, with two flying jibs was seen, which to the unaided eye appeared about half mast "down," with the hull entirely out of sight. There could be no mistake about this, as the vessel was observed in the north while the sun was shining brightly from the southwest.

Through the mounted telescope of about fifty diameters all of the sails, masts, and hull were plainly visible to the water line. Sloop was seen at the same time as the two-mast schooner, appearing only as a white speck on the horizon. Through the telescope the sail, mast, jib, and hull were visible.

February 11, 4 p. m.; large schooner observed in southwest; appearance to unaided eye about half mast "down," with the hull entirely hidden from view. Through the telescope the lower masts, sails, and hull were visible.

February 13, 9:45 a. m.; sloop observed about 9 miles from shore. With the most careful observation with the eye alone, not more than half of the mainsail could lie seen. As the vessel was in a calm, the very best opportunity was afforded for critical observation. The telescope not only showed plainly the topmast and rigging, but also the hull down to the surface of the water.

February 15, 10:15 a. m.; a jigger rig yacht came into view with only about one third masts visible above horizon. The horizon line was fine; there was no haze, and topsail of the vessel was clear cut and well defined to the naked eye, the water beneath apparently occulting two thirds of sails and all of hull. At first glance it appeared as a white speck on the horizon; the most careful view with the unaided vision would not permit the sight of the lower sails and hull.

The view through the telescope showed such a contrast, through which the hull became visible with marked distinctness, the vessel in the telescopic field being visible down to the water upon which it sailed.

February 17, 10 a. m.; schooner observed in north-west. A sailor passing at the time was asked how far "down" the vessel appeared to him. "About half mast 'down,'" he said. With the axis of the telescope about 5 feet above water level, the hull was visible, and was observed by members of our Staff and Corps, as well as by the sailor.

The telescope was then taken to the water's edge, with the tripod lying on the sand and the telescope resting on a small support, so that it was about 15 inches above the Gulf level. Through the instrument in this position, with the head on the sand at the subjective eyepiece, the hull was still visible.

The sailor considered the vessel about 8 miles distant. With the eye 15 inches above the water, the horizon, if the earth were convex, would be about 1½ miles distant, leaving 6½ miles for declination beyond the horizon, which would. amount, according to the ratio of curvature, to a little over 30 feet, placing the hull that far below the line of vision!

Sanibel Light Visible 34 Miles

The lighthouse on Sanibel Island is 34 miles N. N. W. from Naples. The light has an elevation of 98 feet above mean tide level. In order that other lights may not be mistaken for lights in lighthouses, every such light possesses certain characteristics by which it may be recognized by all--such as alternating or intermittent flashes, different colors, etc.

On the evening of January 5, 1897, from the dock extending into the Gulf at Naples, the Sanibel Light was observed through the large mounted telescope, directed to N. N. W. The intermitting flashes showed it to be the Sanibel Light beyond any reasonable doubt.

Let us consider the utter impossibility of observing this light at this distance if the earth were convex. The axis of the telescope was about 17 feet above the water level; this would place the horizon about 5 miles away, leaving 29 miles of declination from the horizon to the lighthouse, which would amount to 560 feet, the required height of the light to be seen from Naples, at an elevation of 17 feet.

Here is manifest a difference of 462 feet between the fallacious Copernican theory and facts of actual observation! This light has also been seen from the Naples dock with the naked eye, under extremely favorable circumstances, concerning which we append the following statement:

In March, 1895, one evening between 8 and 9 o'clock p.m., I, in company with Mr. Drummond and Mr. Hugh McDonald, of Covington, Ky., and Thos. E. Hart and N. Walker, of Marco, Fla., saw from the pier or dock at Naples, Fla., the Sanibel Light in lighthouse on Sanibel Island, N. N. W. from Naples. The evening was clear and the light shone clearly.

The light is an intermittent one, with one bright flash and two less bright; these flashes came in regular order throughout the time of observation, so that we could not have been mistaken regarding it being the Sanibel Light. It was at low tide, Gulf very smooth, with northeast wind for several days previous. The mean difference between high and low tides here is about 3½ feet. The heighth of the floor of the pier from low tide is about 12 feet.--David N. Walker, Sailor, Marco, Fla.

I was present at the observation referred to, and attest the truthfulness of the above statements.--N. Walker, Marco, Fla.

Cape Romano Visible 25 Miles With the Eye at Water Level

With clear atmosphere and calm weather, the distance at which objects can be seen upon the sea is greater than would be possible upon the basis of the earth's convexity. We append the following statement handed to us by a citizen of Marco, Fla., an old resident, familiar with every point along the Florida west coast:

About the 29th of January, 1895, at about 4 p. in., Mr. S. E. Williams and myself, from Rabbit Key, a small island just north of Pavilion Key, and a little south of Chokoloskee Pass, observed Cape Romano at a distance of about 25 miles. The timber on the cape was as plain to the unaided eye as if it had been only a few miles away with ordinary atmosphere. A little schooner yacht that had passed us and had been out of sight for over two hours, was in plain view, even to her hull.

There was also a schooner that we had not seen before, sailing along the channel from Coon Key to Cape Romano; but I do not remember whether we could see her hull or not. The distance I should judge to be about 25 miles. I believe that we could have seen the above-named objects 10 miles farther, as we laid down over the deck of the boat, with our heads on a level with the water, and we could see the cape, schooners, etc., as plainly as when on the cabin.

The sky was cloudy, and we could not see the sun. There was very little breeze at the time; what there was, came from the south. The reason I know that it was Cape Romano, is that there is no other land W. N. W. from Rabbit Key--the course.

Also, Mr. N. Walker, of Marco, and Robert Anderson, of Hotel Naples, saw Sanibel Light from the Naples dock one night in March of the same year. I would make affidavit to the above, except as to distance, which may not be exactly correct.--Thos. E. Hart, Marco, Fla.

OBJECTIONS BY ILLOGICAL CRITICS

Observations on the Gulf of Mexico by the Koreshan Geodetic Staff

THE OBJECTION had been so often urged by illogical critics that tests upon inland waters were not satisfactory, that it was decided that observations be made upon the Gulf itself, the conformity of which to the contour of the earth no sane mind will question. Against the results of such observations, no subterfuge can be brought to bear. For this reason also, the air line was surveyed upon the Gulf coast.

Six and one half miles lie between the points of the mainland extending into the Gulf at Gordon's and Doctor's Passes. These points are long sand-bars, the elevation of which is equal to the high tide of the Gulf. On the point at Doctor's Pass a large target 3½ feet square was fixed upon supports; the top of the target was just 5 feet above low tide.

On March 2, the mounted telescope was taken to Gordon's Pass, and the visual axis of the instrument was fixed at an elevation of 3½ feet above low tide. At this altitude above the low tide level, all of the surface of the target was visible, and the white line of the sandy beach lying beneath it was distinct. No convexity was observable at this elevation.

On the morning of March 3, at a time when the Gulf was calm, the observation was repeated. With the telescope fixed 2 feet above the water level, the target was still visible; the same at 18 inches, and finally by reclining at the water's edge, with the axis of the instrument 12 inches above the water's surface, the target was still in view.

Under the conditions of the last observation, if the water were convex the horizon would be only 1¼ miles distant, leaving 5¼ miles of the surface of the Gulf to decline downward,--amounting to 18 1-3 feet. As the target was only 5 feet above the low tide water level, it would be 13 1-3 feet below the line of vision. After deducting nearly one seventh of this declination to make up for the usual allowance for refraction, 11 1-3 feet would remain as the amount of depression of the target below a line extending through the visual axis of the telescope over the horizon, to the distance of 5½ miles beyond the horizon. All of these observations were repeated in the afternoon, with the same results.

Experiments on Naples Bay

A straight reach of 4½ miles was found upon the smooth waters of Naples Bay. At the most southern extremity a target of white cloth 29x30 inches was fixed upon an upright with cross-arms; the top of the target stood 2 feet above the high-tide mark, leaving a space of 4 or 5 inches to the water's surface.

On March 5, at time of high tide, the Staff sailed to the farthest point northeast from which the target could be seen with the telescope. To the naked eye, the target was entirely invisible. The horizon seemed to occult the lower limbs of the belt of mangrove trees constituting the background of the view.

Over the water at the point of observation, the telescope was fixed at an altitude of 30 inches above the water, and through it the target stood out in bold relief. The instrument was then lowered to within 18 inches, with the same observed results. Afterward, at the height of 10 inches above the water, the entire surface of the target was still visible.

Very careful observations were made and repeated with the telescope at this altitude. The target was clear cut and well defined, and even the space between the bottom of the target and the water was observable. Then, to make the test absolutely satisfactory and conclusive, the telescope was fixed upon the water's surface; with the instrument almost touching the water--indeed, it could not be placed closer without wetting the lenses--long and careful observations were made. There could be no mistake; the entire surface of the target could be seen, with a small dark line of the background appearing beneath it.

The terrestrial eyepiece was then exchanged for the astronomical eyepiece of greater power. The target was increased in size, and the relations of the target and the water's surface and the background came out still more noticeably. The object glass is 3 inches in diameter; the axis of the telescope was 2 inches above the water. On the basis of convexity, the horizon would be but one half mile away--for the declination for one half mile is considered to be 2 inches--leaving 4 miles of surface to decline from the horizon point, amounting to 10¾ feet. The target would have to be higher than 10¾ feet above the water in order to be seen; as it was at an altitude of only two feet, it would be 8¾ feet below the line of sight.

Water's Concavity Visible

These are the most satisfactory observations thus far made by the Geodetic Staff, because the tests were more crucial. The results were conclusive, as they afforded an ocular demonstration of the earth's concavity. A stake 2 feet in height was placed midway between the Observing Station and the target, with cross-bar at top of stake.

With the telescope at the same altitude, the cross-bar was observed to be a little below the top of the tar-get, with the target foreshortened by perspective to a breadth equal to about one half the length of the stake. With the visual axis of the telescope 2 inches above the water, the cross-bar was seen to be in line with the top of the target.

Besides this observation, an absolutely satisfactory view was had of the water surface itself. With the telescope placed absolutely level, the water appeared to slope gradually upward to the center of the telescopic field. With the objective end of the telescope placed a little upward from the true level, and with the water still visible near the objective end of the instrument, the actual concavity of the water--a mid-way depression--was clearly observable.

This midway depression was at the point of the stake with cross-bar midway between the point of observation and the target, from which midway depression there was a gradual slope upward to the target. This view was obtained by the long, terrestrial eye-piece, and also by the astronomical eyepiece, the concavity through the latter being the more marked. There could be no mistake as to the concave arc; the. water was seen to be not convex; it did not appear to be a plane, but concave!

PREPARATIONS FOR THE GEODETIC SURVEY

FOR SEVERAL WEEKS previous to the beginning of the geodetic operations, the numerous tourists from the north and from Europe, and residents of Naples, standing upon Col. Haldeman's long dock, saw 2 15-foot, 2 x 6 inch perpendicular stakes outlined against the southern horizon. They marked historic points along the line of the first survey that determined the true contour and ratio of curvature of the earth's surface.

From the fixed stake on the approach to the Naples dock, the stakes marked the direction of the meridian line. Standing in a long line like sentinels, were the lesser stakes that indicated shorter intervals of space. We had conducted a coast survey; with surveyor's instruments had measured a line or path along which the Rectilineator was to be moved section by section in precise adjustments, and eighths of miles marked by stakes, for 4½ miles down the coast.

As the air line was to be straight, and as the shore line was a little irregular, the land elevation above the water level varied from 3 to 5 feet. Excavations were necessary, and much other work of similar character, to remove all obstructions and clear the way for convenient and uninterrupted operations when the adjustments began. We refer to these incidental preparations, in connection with all other factors involved in obtaining the results, to give something of an idea of the magnitude of the undertaking we had before us.

It will show that care was taken to attain accuracy; that we were faithful and persistent in the execution of our plans; that we understood what was required to determine the facts involved in the question,--geometrically, mathematically, practically, and mechanically; to manifest to the reader that we faithfully detail the entire proceeding--all of the obstacles and difficulties, and how they were removed, as well as demonstration of principles and facts of measurements.

It proved not only that the survey was made, and made honestly, and success achieved, but also to show that in consideration of the fact that it was the first attempt in the history of the world to make such a survey, its accomplishment is a marvel!

LEVELS, PLUMBS, APPURTENANCES, AND RECORDS

INASMUCH as the Geodetic Survey was extended through space by means of right angles, regardless of any other method of determination of a straight line, and regardless of the consequences, it is obvious that it was not extended by any leveling process.

By reference to cut No. 4, Plate 1, it will be seen that the rectiline would vary from the water level in ever-increasing angles from the beginning to the end of the line. If the earth were convex, the line at the end of 4 miles would be higher than at the beginning, and the angles would be divergent from the beginning; if concave, convergent from the beginning. We used levels for two purposes: First, to level the first section; second, to ascertain and record the variation of the sections from the water horizontal at given points along the coast.

By reference to the "Comprehensive View of the Air Line" (diagram 4, Plate 1), the reader will understand how the plumbline should hang with reference to the right-angled bars, first, if the earth were convex; second, if it were flat; and third, on the basis of the concavity.

The leveling of the first section was the point for the exercise and application of the greatest skill and accuracy; the first section must be accurately leveled. For this purpose we applied one of the finest and most sensitive spirit levels obtainable. In connection with this we had our 12-foot mercurial geodetic level, invented specially for this survey. Being 12 feet in length, it was susceptible of being used with great accuracy and precision.

Applied to the first section, the spirit and mercurial levels agreed. The plumb was also applied to the cross-arms of the first section as additional corroboration. The horizon was also observed in relation to the long straight-edge formed by a number of adjustments, and the straight-edge was perfectly parallel with the clear-cut water line of the Gulf of Mexico, viewed from a point three or four rods back of the apparatus, so as to place the under edge of the straight-edge and the water line in apparent contiguity.

The leveling was a careful, painstaking, and successful work, witnessed by every member of the Staff, and finally pronounced perfect at 8:50 on the morning of March 18, 1897. From thence the line was projected on the basis of the principles which we have demonstrated.

A convenient chest was moved along the line as the work progressed, with thermometer, microscope, calipers, rules, compass, spirit level, triangles, pro-tractor, telescope, thumb bolts, adjusting gauges, celluloid test card, etc., and the books of the Staff for the purpose of making the most accurate observations and measurements, and recording the same on the field of operations in the presence of all the witnesses.

Every item of adjustment, test, observation, and measurement was checked in the check record book, and described in detail in the daily record book, to which are appended the signatures of all operators and witnesses. The facts of preparation, measurements, and survey contained in this work are taken from the records, attested and sworn to by the entire Geodetic Staff and the investigating committee.

Personnel of the Staff, Investigating Committee, and Corps of Witnesses

In our line of argument it is necessary to intro-duce the operators and witnesses, that the reader may judge of the character of the testimony concerning the facts observed; and to this end we publish the names of all those connected in any way with the experiments and survey conducted on the Florida coast. The operations and observations were not witnessed by the Operating Staff alone.

Appended to this work are the statements of the visiting and investigating committee, concerning the facts observed when the air line was projected into the water on May 5, 1897, and the repetition of the same on May 8; also the sworn statements of the operators and watchmen concerning the precautions taken to prevent any one tampering with the apparatus or its adjustments. In the list of the Operating Staff we briefly mention the position each occupied, and the class of work to which each was assigned:

U. G. Morrow, Geodesist, inventor of the Rectilineator; in charge of field operations, experiments, and observations; tested adjustments and measurements, and checked same in Record Books.

*L. M. Boomer, General Manager.

Rev. E. M. Castle, of the University System of the Koreshan Unity; inventor of the System of Reversals of sections of the Rectilineator.

G. T. Ordway, Operator; manipulated set screw No. 1; detached each rear section, and transferred same for forward adjustment; made reversals in accordance with the formula of the Castle System; signaled tide measures from stationary caisson.

J. J. Williamson, Assistant Operator; manipulated set screw No. 2; assisted in detachment of each rear section, and in the reversals; watchman.

*H. B. Boomer, Secretary.

George W. Hunt, Engineer; directed emplacement of the 8-foot platformed standards, and adjustment of castings which received the sections of the Rectilineator; in charge of all excavations, and setting of Tide Staffs.

P. W. Campbell, Mechanic and First Assistant Engineer; Assistant Watchman.

Allen H. Andrews, Second Assistant Engineer; Assistant Watchman.

Corps of Staff Assistants

Gustave Faber	Leroy L'Amoreaux
Charles Mealy	Laurence Bubbett

Visiting and Investigating Committee

Victoria Gratia, Pre-Eminent of the Koreshan Unity.
Rev. E. M. Castle, of the Koreshan University, Estero, Fla.
Prof. O. F. L'Amoreaux, A. M., Ph. D., Estero, Fla.
C. Sterling Baldwin, M. D.
Mrs. Ada Welton.
T. R. Ehney, Postmaster at Naples, Fla.
W. D. Puerifoy, Naples, Fla.
S. L. Green, M. D., Marco, Fla.

Other Witnesses and Visitors

Hugh McDonald, Covington, Ky.
Mrs. Hugh McDonald, Covington, Ky.
Miss Ann Haldeman, Louisville, Ky.
Miss Lucy Lemon, Louisville, Ky.
Miss Elsie Frederickson, Louisville, Ky.
J. T. Smith, Springfield, Ill.
Mr. Strauss, of Louisville Courier-Journal.

Capt. Robert Gilbert, Estero, Fla.
Richard Gilbert, Punta Rassa, Fla.
Mrs. Elizabeth Robinson, Chicago, Ill.
R. B. Gilbert, Punta Rassa, Fla.
Mrs. Esther Stotler, Estero, Fla.
Miss Rose Welton, Estero, Fla.
Carl Luettich, Estero, Fla.
Lester Wintersgill, Estero, Fla.
G. R. Calhoun, Plant City, Fla.
Thos. E. Hart, Marco, Fla.
D. N. Walker, Marco, Fla.
N. Walker, Marco, Fla.
Miss K. M. Large, Naples, Fla.
Neal Harris, Marco, Fla.

*Called to Chicago by telegram announcing the death of their father, Mr. L. S. Boomer, before Survey began; assisted in preparations.

RESULTS AND INEVITABLE CONCLUSIONS

Details of Measurements and Extension of the Air Line into the Water

WHEN we suspended the plumbline at the first adjustment of the Geodetic Apparatus, we established beyond all doubt the direction of the earth's radius, or the perpendicular at the initial station.

Poised upon the pivot of adjustment, the bubble in the graduated vial of the spirit level measured equal distances from the central division of the scale. The mercury in the 12-foot mercurial geodetic level stood at equal altitudes in the perpendicular tubes, in demonstration of the fact that the level is at right angles with the perpendicular radius of the earth.

The plumb and level invariably tell the truth; they are silent witnesses testifying from the standpoint of unseen "energies," which man cannot bribe nor change to suit a theory. The bubble-shifts at the various stations throughout the line of survey, whether corroborating or denying preconceived opinions, must be accepted as conclusive.

With perpendicular and horizontal definitely fixed at the starting point in our survey and in our argument concerning the evidences afforded in the line projected, we have factors which constitute an indisputable basis of reference. Once leveled, the direction of our line was fixed, from which it was not possible to depart; the bolts which held together the brass facings on the adjusted right-angled cross-arms would admit of no change.

The very principles of construction of the apparatus compelled the maintenance of the rectiline from the beginning to the end. The line projected must terminate somewhere, either in space or in the water, according as the earth would be found to be convex or concave.

If the earth were convex, the line would extend into space, as before explained; as the line would proceed, the bubble in the spirit level would shift at each successive application, more and more toward the south from the central division of the scale, while the plumbline hanging in the direction of the perpendicular, or the earth's radii at the various stations, would hang toward the initial station. If concave, the conditions and positions of the levels and plumb would be the reverse of those on a convex surface; if flat, they would be the same continually, as at the beginning of the line.

By reference to Cut 4, Plate 1, the relations of the plumb to an extension of the horizontal at the initial station may be clearly seen, as regards the convex, the flat, and the concave theories. In conjunction with the tests of levels and plumb, the observations of the Gulf horizon were made, as before explained.

At the beginning of the line, the straight-edges of the apparatus when in adjustment were parallel with the horizon. On a convex arc, the straight-edges and the horizon line would appear to converge toward the north with increasing angle, as the line proceeded; if flat, their original parallel relations would be apparent throughout the line; and if concave, the apparent convergence would be toward the south, or in the direction of the movement of the apparatus.

The Testimony of Levels, Plumbs, and Horizon

We present the evidences of the readings of the levels, plumb, and horizon, because the evidences afforded by these means are independent of any measurements of altitude of the line surveyed above the mean tide level; we offer them as corroborative of the measurements obtained.

It will be found upon comparison with the table of measurements in this chapter, that these evidences are in harmony with the measurements and facts which constitute the factors of our direct demonstration.

The spirit level was applied in test of position of sections at every twelfth adjustment throughout the line. For the first several tests, the divergence of the water line and the air line was too slight to be detected by means of the level; and it was not until near the end of the first eighth-of-mile division of the line that any difference was thus manifest. The bubble had shifted a little--toward the north, or rear section of the apparatus. From the first point of the manifest deviation until the end of the line, the angle increased proportionately to the distance traversed.

This was corroborated also by the position of the plumbline, and the observed increase of angle between the straight-edges and horizon, always converging toward the south.

We have thus far referred to these angles in general terms; the question would arise, What were the actual angle measurements as ascertained by the levels, plumb, and horizon, at points where all these tests were applied? If there were variations, how great or small were the variations?

The divisions on the graduated scale of the spirit level were .075 of an inch apart; the plumb was suspended from top of the 4-foot right-angled cross-arms, and the angle read on plate at the bottom of the cross-arms; the 12-foot mercurial level determined the angle for 12 feet, while the observed horizon was related to straight-edges 36 feet in length, and therefore determined the angle for 36 feet.

We give below the results observed at the end of the first mile, the second mile, and at end of 2⅜ miles, at last adjustment of the apparatus in the southerly direction.

Spirit Level, shift of bubble toward north end of the vial, as measured on the graduated scale:

1 mi., .0375 in.; 2 miles, .077 in.; 2⅜ mi., .089 in.

Plumbline, measurement on arc of 4 feet radius, as related to right-angled cross-arms:

1 mi., .015 in.; 2 mi., .037 in.; 2⅜ mi., .044 in.

Mercurial geodetic level, indicating angle of divergence of air line and horizontal at points of test, for the space of 12 feet:

1 mi., .042 in.; 2 mi., .094 in.; 2⅜ mi., .115 in.

The Horizon, indicating angle for space of 36 feet, as accurately as could be measured with the eye at a distance of 15 feet from the apparatus:

1 mi., .15 in.; 2 mi., .34 in.; 2⅜ mi., .51 in.

These readings, taken as a basis of mathematical calculations, will be found to very nearly conform to the relations of chord and radii, over an arc of 25,000 miles circumference, and are evidences of the angles increased in about the proper ratio, as the surveyed line progressed from the beginning to the end. The chord, tangibly constructed, was constantly converging with the arc, as we have shown by the four independent sources of evidence; and we now purpose presenting such a network of facts of absolute and direct demonstration as to render our position invulnerable, and our premise impregnable.

Details of Visual Projection of Air Line Over Gordon's Pass

From the Naples dock looking south, a white line in the Florida sand marks the line of excavations for emplacement of the standards of the Rectilineator, extending to Gordon's Pass, in the southern horizon. This white line is the actual mark we have made in the world,--our path to success, the route of demonstration. Carefully and patiently for eight weeks we followed this course, making precise adjustments,--painstaking, careful, tedious, and trying work, by which the facts we are publishing were obtained.

We diagram the Air Line, the coast line, and the water's surface to the extremity of the line, covering the space of 4⅛ miles. A is the point of beginning of the line at altitude of 128 inches; B is the point where the rectiline extended into the waters of the Gulf of Mexico. The coast line can be easily traced without lettering; C is Gordon's Pass; D, the long sand-bar extending into the Gulf, through an excavation in which the Air Line was projected; the excavation admitted of vision of the water horizon from the 2⅜ mile station, where the final adjustments of the apparatus were made.

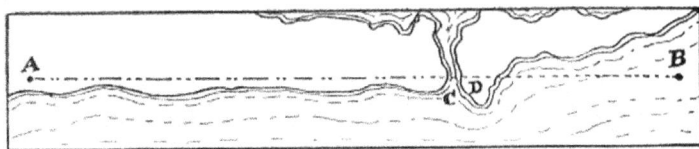

Map of Coast Line and Gordon's Pass, Showing Course of Air Line and Point of Projection Into the Gulf

All of the definite measurements of altitude of the line and of the minute angles of divergence were made with reference to the position of the apparatus from the beginning of the line to the end of the adjustments, covering a distance of 2⅜ miles; the line was tangibly built for this distance. The cross-arms extended 2 feet below the horizontal axis or middle line of the sections, and at the end of 2⅜ miles were within 7 inches of the ground; the axis of the apparatus was rapidly converging with the water level.

We had manipulated the apparatus as far as practicable and possible, within the limits of our conveniences for operation from the starting point of 128-inch altitude. The Air Line at this point was 54 inches nearer the water's surface than at the beginning; whereas, if the earth were convex, the line would have been 54 inches above the vertical point of the original 128 inches, making a difference of 108 inches, or 9 feet, in the position of the apparatus from that really obtained.

Extension of Air Line into the Water

For the reasons given above, the extension of the Air Line into the water, or a continuance of the line in any direction, we had to employ another method of survey. As the line would extend across the Pass and over the sand-bar, through the excavation, and into the Gulf south of the Pass, it was necessary to finish the line by a visual process.

In order that this might be done as accurately as possible, obviating any errors that might arise from the adjustment of such an apparatus as the engineer's level, parallel with the horizontal axis of the apparatus, two points on the Air Line surveyed by the apparatus, one-eighth of a mile apart, were taken as tile sighting stations; these points were tide staffs Nos. 19 and 20.

On these staffs we had left the record of the altitude of the Air Line. The large telescope provided with horizontal cross-hair, was adjusted at staff No. 19, so that its line of collimation was at the same altitude as the surveyed line at that point. On staff No. 20, one-eighth of a mile distant, a steel strip was fixed horizontally at altitude of Air Line at that point, upon which to train the telescope. When the telescope was adjusted so that the cross-hair was in line with the steel strip, the simple matter of projecting the remainder of the line visually is easily comprehended.

At these distances from the beginning of the line, with the tending divergence toward the water, refraction and visual curvilineation would involve so small a factor of departure from the rectiline for the remaining distance as to give an approximately correct reading on the staffs to the end. Through the telescope, the steel strip and the cross-hair were observed to a point below the Gulf horizon--the visual line extending into the water south of the Pass.

The following diagram gives a perpendicular view of the line, showing the land elevation. XY represents the arc of curvature; A, the beginning of the survey; B, the point of projection into the water; C, the Pass, and D, the sand-bar through which excavation was made; E, tide staff No. 19, with telescope; F, tide staff No. 20. The continuous line is the line surveyed by section adjustments; the dotted line is the portion of the Air Line projected visually.

Profile View of Land Elevation, Showing Process of Visual Projection of Air Line Into the Gulf

For the purposes of measurement and calculation of ratio of curvature, it was necessary to locate the point on the Gulf where the line extended into the water. This was done by directing our sail-boat beyond the Pass, in line with the telescope axis. When the lower part of the hull appeared just above the cross-hair, it was obvious that the point was marked.

By means of our signal code, the observer at the telescope transmitted the information that the point was reached by the sailors in the boat, and the occupants on getting signal, went directly ashore, and consulting stakes, gave the distance as approximately 4⅓ miles from beginning of survey.

These observations were participated in by the visiting and investigating committee, as well as the Operating Staff, who conducted a repetition of the observations three days later.

At the time of the observations, sketches of the telescopic view were made, showing the boat on the Gulf where the Air Line was projected into the water, as the converging chord of arc; and for the benefit of the reader we herewith produce sketch, showing the cross-hair in telescope at tide staff No. 19, the picture of tide staff No. 20, with steel strip attached in line with the cross-hair.

The visual line connecting the same, extending beyond and projecting into the water, completed the experiment, the results having been proclaimed by the Founder of the Koreshan Cosmogony twenty-seven years previous to the demonstration on the Florida west coast.

For the purpose of further testing the apparatus, we retraced the surveyed line by means of the apparatus for the distance of ⅜ mile, taking the last forward adjustment as the first adjustment on the return survey. If the hair-line of the apparatus returned to the same point on the tide staffs, it would be a further demonstration of the accuracy of our work. An error of deviation from the rectiline would be applied on the return, and would consequently be made manifest by having fixed points as bases of reference, such as were recorded upon the tide staffs on the forward survey.

Upon return to the staffs the hair-line of the apparatus fell upon nearly the same points, as per the figures in the table of measurements under the following subhead, demonstrating the remarkable accuracy and efficiency of the apparatus employed to make the first direct test of the earth's surface in the history of the world.

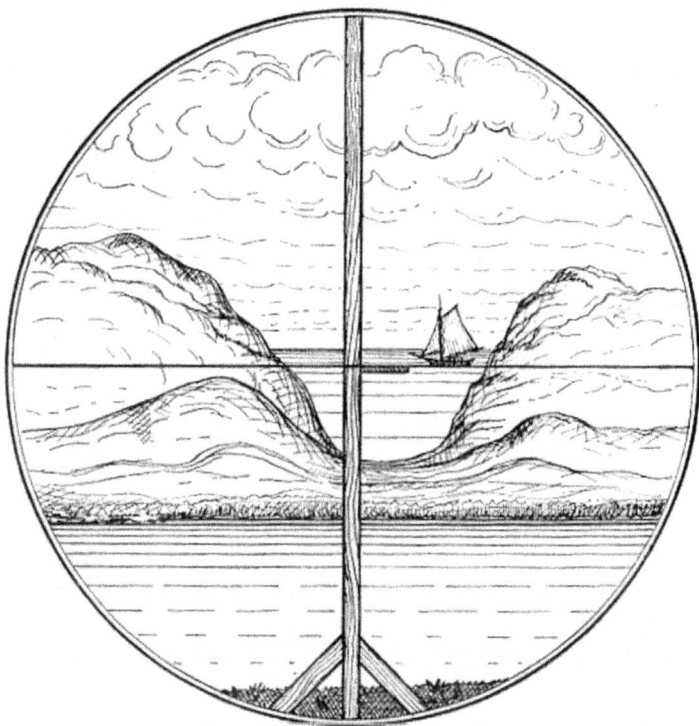

Sketch of the Telescopic View
Showing Extension of the Air Line Into the Gulf Through Excavation South of Gordon's Pass;
also Relation of Cross-hair and Steel Strip to Horizon and Boat in the Distance

Dates of Measurements on Tide Staffs.	Distance in Feet From Beginning.	Distance in Miles.	Number of Adjustments.	Number of Tide Staff on Beach.	Inches Altitude of Air Line Above Fixed Datum Line.	Distance of Air Line Below Secondary Datum Line, inches.	Calculated Ratio of Concave Curvature, inches.	Difference Between the Ratios, inches.
March 18	0	0	0	1	128	0.	0.	0.
" 19	660	1/8	55	2	127.85	.15	.125	.025
" 23	1,320	1/4	110	3	127.74	.26	.5	.24
" 24	1,980	3/8	165	4	126.625	1.375	1.125	.25
" 25	2,640	1/2	220	5	126.125	1.875	2.	.125
" 27	3,300	5/8	275	6	124.125	3.875	3.125	.75
" 30	3,960	3/4	330	7	123.675	4.375	4.5	.125
" 31	4,620	7/8	385	8	121.57	6.43	6.125	.305
April 1	5,280	1	440	9	119.98	8.02	8.	.02
" 2	5,940	1 1/8	495	10	117.875	10.125	10.125	.0
" 8	6,600	1 1/4	550	11	116.44	11.56	12.5	.94
" 9	7,260	1 3/8	605	12	113.69	14.31	15.125	.815
" 13	7,920	1 1/2	660	13	111.07	16.93	18.	1.07
" 14	8,580	1 5/8	715	14	107.19	20.81	21.125	.315
" 14	9,240	1 3/4	770	15	104.69	23.31	24.5	1.19
" 15	9,900	1 7/8	825	16	101.69	26.31	28.125	1.825
" 16	10,560	2	880	17	97.38	30.62	32.	1.38
" 24	11,220	2 1/8	935	18	93.44	34.56	36.125	1.565
" 26	11,880	2 1/4	990	19	85.32	42.68	40.5	2.18
" 27	12,540	2 3/8	1,045	20	79.75	48.25	45.125	3 125
May 8	13,200	2 1/2		21	74.	54.	50.	4.
" 8	13,860	2 5/8		22	68.	60.	55.125	4.875
" 8	14,520	2 3/4		23	63.	65.	60.5	4.5
" 8	15,840	3		24	53.	75.	72.	3.
" 8	21,780	4 1/8		25	0.	128.	136.125	8.125
RETURN SURVEY.								
" 6	12,540	2 3/8	1,084	20	79.75	48.25	45.125	3.125
" 11	11,880	2 1/4	1,140	19	85.47	42.53	40.5	2.03
" 11	11,220	2 1/8	1,194	18	93.68	34.32	36.125	1.805
" 11	10,560	2	1,250	17	97.13	30.87	32.	1.13

TABLE SHOWING ALTITUDE OF AIR LINE ABOVE DATUM LINE AT EVERY STATION OF
SURVEY, WITH THE MEASURES COMPARED WITH THE CALCULATED CURVATURE
SKETCH OF THE TELESCOPIC VIEW

Measurement of Altitude of Air Line on 25 Tide Staffs

As the Rectilineator was moved forward, section by section, in the direction of the tide staffs, the relation of the hair-line of the sections to the 128-inch altitude, or secondary datum line, was easily obtained by measurement. At no place throughout the line of survey did the sections rise above the original 128-inch altitude.

At the end of the first 660 feet, where the calculated ratio would indicate .125 of an inch curvation, the hair-line of the apparatus fell .15 of an inch below the 128-inch altitude, being a difference of only .025 of an inch. This is in easy contrast with the conditions that would result if the earth curved downward instead of upward. If it were convex, the hair-line of the sections would have fallen about .125 of an inch above the 128-inch altitude. The tide staffs marked brilliant points of interest throughout the survey; for each measurement was a demonstration not only of the fact that the earth curved toward its chord, but also the ratio of its concavity.

For easy reference and definite record, we have condensed all of the facts of measurements of the Koreshan Geodetic Survey into the preceding table of distances and altitudes, by means of which the results of our work may be easily compared with the calculated ratio, and how nearly, at each tide staff, the hair-line of the apparatus came to indicating the ratio which we have since calculated, and here present for comparison and study.

It will be noticed that the difference between the measured and. the calculated ratios increases toward the end of the survey; the apparent rapid approach of the line to the water's surface after 2¼ miles had been surveyed is due, not to actual approach to the mean water level of even curvature, nor to inaccurate work or measurements of the tide levels, but to the crowding of the waters around the mouth of the Pass, creating a slight irregularity in the water level in that vicinity.

It will also be noticed that the line projected into (the water at the distance of approximately 4¼ miles instead of 4 miles, which is due to refraction and incurvation of the visual lines from the 2⅜ miles station to the water's surface beyond the Pass.

Comparative Results of Altitude Measurements on Concave and Convex Surfaces

It may be asked, What would be the facts of measurement if such survey as ours were made upon a convex surface? We append table of comparative results for reference, study, or test by calculation. Having two parallel arcs; the mean tide level, and the secondary datum line, connected by staffs of 128 inches altitude as fixed bases of reference, it may be seen that on a convex earth, instead of the air line falling below the vertical point of every staff, it would rise above, and increase in excess of altitude above the original 128 inches as to the square of the distance.

For instance; at the end of the first mile, instead of the surveyed line falling 8.02 inches below the secondary datum line, it would rise about as far above, making a difference of about 16 inches in the altitude of the line on the concave, and the line on the convex earth; while at the end of 4⅛ miles, instead of the line extending into the water, the line would be projected out into space, 136.125 inches above the original 128 inches, making a difference between the two theories of 264.125 inches, or about 22 feet, in 4⅛ miles.

Miles Distance From Beginning.	Altitude Surveyed Line, inches.	Distance Below 128-Inch Altitude on Concave Surface, inches.	Calculated Distance above 128-Inch Altitude on Convex Surface, inches.	Difference Between Concave and Convex Readings, inches.	Total Altitude of Tangent to Mean Tide, Convex Surface, inches.	Total Altitude in Feet.
0	128.	0.	0.	0.	128.	10.66
⅛	127.85	.15	.125	.275	128.125	10.67
¼	127.74	.26	.5	.76	128.5	10.7
⅜	126.625	1.375	1.125	2.49	129.125	10.76
½	126.125	1.875	2.	3.875	130.	10.83
⅝	124.125	3.875	3.125	6.99	131.125	10.94
¾	123.675	4.375	4.5	8.875	132.5	11.04
⅞	121.57	6.43	6.125	12.555	134.125	11.17
1	119.98	8.02	8.	16.02	136.	11.33
1⅛	117.875	10.125	10.125	20.25	138.125	11.5
1¼	116.44	11.56	12.5	24.06	140.5	11.7
1⅜	113.69	14.31	15.125	29.435	143.125	11.92
1½	111.07	16.93	18.	34.93	146.	12.16
1⅝	107.19	20.81	21.125	41.935	149.125	12.42
1¾	104.69	23.31	24.5	47.81	152.5	12.7
1⅞	101.69	26.31	28.125	54.435	156.125	13.01
2	97.38	30.62	32.	62.62	160.	13.3
2⅛	93.44	34.56	36.125	70.685	164.125	13.67
2¼	85.32	42.68	40.5	83.18	168.5	14.04
2⅜	79.75	48.25	45.125	93.375	173.125	14.42
2½	74.	54.	50.	104.	178.	14.8
2⅝	68.	60.	55.125	115.125	183.125	15.26
2¾	63.	65.	60.5	125.	185.5	15.7
3	53.	75.	72.	147.	200.	16.6
4⅛	0.	128.	136.125	264.125	264.125	22.01

TABLE CONTRASTING AIR LINE AS SURVEYED, WITH CALCULATED RESULTS UPON CONVEX SURFACE

If the reader will study the construction of the Rectilineator and the methods used in operating same, it will be clearly understood that we did not make an error of 22 feet in running 4⅛ miles.

In presenting the foregoing line of experiments, both optical and mechanical, we are giving to the world the benefit of much painstaking effort in deter-mining facts as to the real contour of the surface of the earth upon which we live.

In conducting the different lines of investigation presented, we have tried to be honest with ourselves and those who investigate our work as detailed. The conclusions reached are irrefutable. The character, honesty of purpose, and ability of those who conducted these experiments we willingly leave to the decision of the reader.

ADDENDA

CAUSE OF VARIATION OF THE PENDULUM

KORESH

IT IS CLAIMED that the surface of the earth moves from west to east nearly twenty-five thousand miles in twenty-four hours. This is a solid body moving through space, according to modern science, at a rotary speed of over one thousand miles an hour, or more than sixteen miles per minute.

Can any sane man imagine that a solid body with this rate of speed, surrounded by a thin atmosphere, could so carry its atmosphere with its momentum as not to produce a contrary motion of the atmosphere, while at the same time it would cause the rotation of an oscillating pendulum?

No one disputes the fact of the motion of the pendulum as first observed by Foucault, and later experimented with by Flammarion. We might, however, question the uniform direction of the oscillating pendulum in a series of experiments. But allowing the experiment to be fair and the motion as reported, we would inquire, What causes such a phenomenon?

Whatsoever the cause of the motion it must be considered with regard to two propositions. The first, the supposition that the earth revolves because the heavens appear to revolve; and the motion of the pendulum is taken as corroborative testimony to an hypothesis, a guess, which still hangs in doubt with the astronomers, for the reason that with howsoever much reinforcement you sustain a guess, it still remains hypothetical; and the astronomers will not stop seeking for still further corroboration because they are still in doubt. We wish to assure our readers that the problem is not settled under the Copernican system.

The second proposition is, that the Koreshan Geodetic Survey has settled forever the fact that the earth is a concave shell, and that man inhabits the cellular sphere. If it could be proven that the earth rotates, then the pendulum would act the same on the inner as it would on the outer surface of a ball, were it the motion of the earth that caused it. It could not, therefore, affect the fact of the Cellular Cosmogony in the least.

According to the Koreshan System, the earth is comparatively stationary, and the heavens are moving within the stationary earth. The sun is moving at the rate of about eighteen thousand miles in twenty-four hours. It is sweeping through space with this velocity and radiating its "energies" into the environing shell, in which there is a corresponding magnetic spiral motion.

To this spiral motion of "energy" is due the rotation of the pendulum, and not to the motion of the earth. First, it will be understood that the pendulum is suspended from a support attached solidly to the body of the earth. Second, it will be noticed that the curve of the earth is practically the same at both extremes of the oscillation, the earth moving just as rapidly at one point as at the other. There could not be a calculable commensuration of difference either in time or space, at the two extremities, as to curvation or the time of longitudinal motion.

If the pendulum be swung from north to south and south to north, at the start, it would be subject to the eastward motion of the earth, which, were the theory of the earth's rotate impression in relation to the pendulum ball true, the ball would apparently move toward the west, and with the opposite swing of the pendulum it would swing equally toward the west-- the motion on one side balancing the other. This would be the effect if the phenomenon were the result of the earth's motion.

How would it be when the pendulum rotated around to the east and west points? The earth would be rotating toward the east, the pendulum swinging east and west. The earth is moving, according to the Copernican hypothesis, at the rate of sixteen miles a minute either with or against the ball, while it swings westward, and at the same rate with it when the ball swings eastward.

If the earth by its rotation affects the motion of the pendulum enough to cause this rotation, why would is not make an appreciable difference distinctively marked, while swinging east and west?

The swinging of a pendulum could bear no possible relation to the earth's rotation, even if the earth were a ball rotating from east to west at the rate of twenty-five thousand miles in twenty-four hours. The question of the relation of the rotating pendulum of Foucault to the rotation of the earth is equal to the question: "If it takes a thousand bundles of shingles to shingle an opera house, how many pancakes will it take to shingle a meeting house?"

The marvelous thing about this experiment, is that any man possessing any claim whatsoever to the title of scientific should accept this solution without asking the question, "May there not be some hypothesis for this motion as reasonable as, or more so than, the hypothesis of the rotation of the earth?"

If a pendulum were swung at the north pole, oscillating laterally over the plane of the earth's rotation, were there such a motion of the earth, the pendulum hung in space (not upon supports fixed solidly in the earth), there would be some sense to the proposition; as it is, it is the veriest nonsense, and later the "scientists" will laugh at their own folly.

THE WONDERFUL PROPERTIES OF RADIUM

(KORESH in FLAMING SWORD, Sept. 18, 1906)

NO MAN will ever understand the character and properties of radium who approaches cautiously and doubtfully the great law of transmutation, the law of the cross, which constitutes the basis of the primitive Christian conception. The discovery, or rather the creation, of the substance called radium has done much to revolutionize the old chemical theory,--the delusion of a hundred years,--and to obliterate the confidence and confusion which have marked the progress of the scholastic hallucination and uncertainty that have revolved around the incompre-hensible atom, which never had an existence but in the hallucinated imagination of densely befuddled and befogged brains.

It has confirmed the physicist in his inscrutable wisdom, founded upon hypothesis, regarding the bombardment and shivering of the atom and its metamorphosis to seven hundred or seventy thousand corpuscular ions and electrons, which also never had an existence except in the deluded brains of experimental explorers of the mysteries of being. It has shivered the timbers of the chemical hypothesis, to substitute for it another more absurd hallucination to be shivered by some other speculator and hypothetical blunderer.

We advise the student of physics not to rely too much upon the hypothetical deviation from the scholastic wisdom that has deluded science for the last seventy years, and which in the mind of the philosopher appeared as an indestructible, nonconvertible, indivisible, and eternal atom.

The empirical creation of radium has dissipated the dreams of the dreamer. Newton's law of gravitation turns out to be no law at all; and the greatness of the discoverer of the so called law of gravitation, in the fallacy of its appeal to the beclouded mentality of the age, becomes mediocre in the comparison of his fallacy with the limelight of its obscuration of the human intellect which it has beclouded. Helmholtz's law of the elevation of temperature proportionate to contraction by gravity, has been bombarded into smithereens through the revelations of radium.

The theory of the contraction of the sun by the application of the frigid zone to the extinct and superannuated vulcanizing activity of the sun, thus increasing the sun's temperature, that is, the increasing of the temperature of the sun by rendering it colder, is becoming an obsolete philosophy in the development of the mysteries of the little composite thing called radium. The chemist's atom has been literally shivered into nothingness.

The sun is no longer in danger of contracting by the loss of its heat through radiation, because of the elevation of its temperature through frigidity. It will no longer grow hot by being colder, through which cold it grows hotter. This absurd humbug offers no more to frighten the world in the vain expectation that in a few billion years the sun will refuse to shine for us.

The little mass of radium has dissipated all of this mental fog for the denser one of the more complicated hypothesis. There is nothing so beatifying to the dreamer after realism as the good working hypotheses that have in the past, and will in the future, hallucinate the visionary. Working hypotheses are like Madam Winslow's soothing syrup--they quiet the, brain and collapse its rational functions and powers for genuine demonstration.

Radium is an incipient star, a miniature sun, an accident so far as its discovery is concerned, for it is the exploitation of experimental, not scientific, investigation. Were its discovery scientific, there would be no mystery in its character and operations to the so called scientific mind. There are no two of the students of its mystery who agree as to the peculiarity of its phenomena; and the deeper the experimenter dives into the vortex of its secrets, the more is he lost in the claustrum of its obscurations.

Never will the "scientific" world fathom its mysteries so long as it obstructs the consciousness (through the materialistic tendencies of "science") from recognizing the counterpart of matter; namely, the sublime and ethereal quality of substance, which has neither the form nor the properties of matter.

Radium is a peculiar character of matter into which many, not all, of the so called elemental forms have entered to create a composite aggregate, so blended into unity as to become homogeneous in quality, thus making a vortex into which the free ethereal essence of space flows and materializes. It is an accidental and experimental stellar nucleus with properties like the stars, but abnormally located because it is the product of accidental and experimental, not scientific research.

It is such a quality of matter as to form a spigot, tapping the essence of ethereal space, thus transmuting the impalpable, non-material essence to its correlate and counterpartal substance. Matter and the correlate spirit of matter are the two qualities of substance that will ultimately be recognized and acknowledged as two distinct qualities of universal substance. Not until such recognition is accorded will there be a condition of the mind properly called scientific.

Science is knowledge, pure and simple; experimental investigation is neither science nor scientific, nor does it result in anything more than the blind gropings of the past, in which one hypothetical air castle has been tumbled down to give place to a more complicated, more doubtful, and more deceptive hallucination.

The phenomena of radium are the product of an incessant metamorphic activity, in which the processes of the materialization of the ethereal essence and the dematerialization of the matter thus partially formed are in constant progress.

Were its properties understood by the "scientific" world, there would be a still greater composite formula in the activities of the material basis of phenomena; and the result would be the tapping of the entire ether with such a resistance of the radiation mechanically contrived, as to create and materialize the entire metallic redundancy of the wealth of the solutions of space.

Ethereal space contains in solution all of the so called elements of matter in substantial but not in material quality. Ether is the product of the processes of metamorphic operations progressing at the two extremities of the cellular space, with intermediate extremities where there are in operation the materializing and dematerializing processes of Nature; hence the perpetuity of ether is maintained through the constant operation of its creation and its correlate rematerialization.

The ether of space, then, is but the product of the change of material substance to its opposite and counterpartal spiritual substance. With the proper spigot this space .can be tapped and its substance materialized at the will of the operator. Its attainment would not be experimental but absolutely scientific. The mechanical possibility would be the result of the application of inventive ingenuity, for it will require mechanical skill to reach the attainment of successfully tapping the universe of space, so as to pour out its metallic and mineral wealth for the uses of man kind.

The world is but now upon the very verge of scientific discovery. Its methods will not be experimental and hypothetical. The time is at hand when the guesswork of the so called scientist will be out of vogue, and there will be substituted the more certain rational processes founded upon the absolute demonstration of the premise from which the reasoner superstructs his fabric.

Hypothesis can never furnish the would-be logician with the foundation for a rational conclusion not as hypothetical as the premise with which he started; nor can hypothesis ever furnish the basis for absolute knowledge. The scientist must not depend upon the primary guess for the solution of the mysteries of being. Guess at nothing. Prove the first demonstration. Let the premise upon which the reason progresses to the solution of the mysteries of life be absolute, not hypothetical; for hypothesis is the basis of all of the air castles which have been builded and overthrown as often as an independent thinker has launched himself upon the mystic deep of speculative navigation.

ASTRONOMICAL MYSTERIES AND HYPOTHESES

(KORESH in FLAMING SWORD, Dec. 12, 1902)

IF HERE is a whole lot of astronomical wisdom (?) being imparted through the Chicago American, written up by Garrett P. Serviss. Mr. Serviss says, referring to a recent astronomical work issued by the West Hendon House Observatory, England, that "The author, Mr. T. W. Back-house, has devoted his time liberally to the observation of phenomena in the heavens, about which the great majority, even of astronomers, know little or nothing. Yet they are phenomena of surpassing and increasing interest, and when they are fully understood they may revolutionize some of the views now entertained concerning the constitution of the universe."

What an admission for a scientific (?) man to make. A possible revolution in the minds of "scientists," men who know--for science means knowledge! If astronomical knowledge may be revolutionized, then the science of modern astronomy hangs upon a very brittle thread. The revolution, however, is coming because the whole Copernican system is predicated upon the basis of assumption, which every astronomer is willing to confess.

Mr. Serviss declares that few discoveries of modern times "affect the imagination with so deep a sense of mystery as does that of the existence of vast invisible masses in the stellar interspaces. Some of these are demonstrably solid bodies of immense magnitude and gravitational power, intimately and inseparably associated with bright stars.

"Others still more strange, are enormously expanded nebulous clouds [which means, literally, cloudy clouds], that radiate not light, but energy ["mere mode of motion"], which, like the Roentgen rays, affect photo-graphic plates and thus render the invisible indirectly visible."

Mr. S. also refers, in his notice of this book, to "a third mysterious form of substance contained in the depths of space, whose presence is manifested by such phenomena as the 'coal sacks' and the 'dark lanes' that are principally in the Milky Way."

All of this impresses us with the vast amount of ignorance which is accumulated in the modern star-gazing mind, and which, from mere modesty, is denominated knowledge--astronomical knowledge. The amount of this kind of wisdom stored and taught is almost as vast as the "illimitable" thing which the "finite," limited mind attempts to comprehend.

The astronomer says the universe is illimitable, therefore incomprehensible, and then sets his mind--which he says is "finite"--at work to comprehend what he denominates the "infinite." The remarkable thing is, that he knows enough to plan the universe without boundaries, making it illimitable, then sets measurable boundaries to his own mind, and does not know enough to comprehend the fact that he cannot comprehend the "infinite" with the "finite," yet still persists in conjuring fables for one generation, which the succeeding generation may discard.

The fact is, the universe is within the boundaries of mental possibility to encompass, but the mind must start right. It cannot assume a premise and reach, by processes of reasoning, anything but the logical deduction; namely, assumption. The premise must be demonstrated to insure a correct conclusion for a rational consecution of argument. The basis of the Copernican system of astronomy is assumption, according to the testimony of Copernicus himself, as herein quoted:

"Neither let any one, so far as hypotheses are concerned, expect anything certain from astronomy, since that science can afford nothing of the kind. The hypothesis [guess] of the terrestrial motion of the earth was nothing but an hypothesis, valuable only so far as it explained phenomena, and not to be considered with reference to absolute truth."

This is an honest confession of ignorance in the mind of the originator of the present system of astronomy. The modern astronomer has simply added an accumulation of guesses, which he has made to fit into the original guess, and which Mr. S. says is threatened with revolution in the mind, providing it does not call a halt upon the attempt of the finite mind to fathom the mysteries of the "infinite."

Finally, Mr. Serviss says: "There is, perhaps, nothing which gives rise to so keen a desire to have knowledge advanced in that particular direction, as does the part of the book relating to the structure of the sidereal universe." That which has no limitation has no structure, because form is the fundamental factor of structure, and limitation is a fundamental property of form.

The material universe has length, breadth, and thickness, or it does not exist. These properties are ever, one of them limitable. What interests Mr. S. the most, are "the 'gaps,' 'rifts,' and 'wisps' in the Milky Way, and the 'radiated systems' and 'flower-like structures' visible amid the infinitely varied array of stars."

Now all this is simplified when we start out with a demonstrated premise. Start with the thing you know, not the thing you guess at, (see the CELLULAR COSMOGONY,) then everything comes easy. The universe is within the comprehension of the mind which, as the offspring of the parent mind, can grow into the amplification of the parent.

A star is the result of a defined focalization of convergent rays of spirit within the luminous ether, which is encompassed by the shell of the universe, and within our own atmosphere.

The nebulæ (clouds) are incomplete focalizations. The "energies" do not merge into a complete focus, therefore the combustion is not so absolute as at the complete focal point; all these manifestations in the stellar space are reflections merely from the shell which comprises the circumference.

FORCE OF VIBRATION APPLICABLE TO AERIAL NAVIGATION

KORESH

IT IS a publicly known fact that the application of the law of vibration to the neutralization of the so called force of gravity, to be utilized for the development of aerial navigation, has been advocated by the Founder of the Koreshan System for years. . .

It matters little to us who takes advantage of the laws of motion to neutralize the direction of motion called gravity, for the purpose of aeronautic navigation. We merely reiterate the statement that aerial ship service will be a success in the near future, but not on the basis of balloons or flying machines. The law of reverse motion will be applied to an airship, cutting the gravic "energy" in two, and electro-magnetic currents will be utilized for the propulsion of the ship. . . .

One of the highest phases of the application of the law of splenic agitation was exhibited when the Lord walked upon the water. The flight of the bird is effected not by the action of the wind or air, but by an action of the brain, producing a vibratory force which buoys the body by an innate "energy." The supreme operation, of this law of vibration has been active in all cases of translation, the specifically recorded instances being those of Enoch, Elijah, and Jesus. These instances may be specifically illustrated by the notable one of the Lord Jesus. The vibration of the atoms of his body was caused by the conspiring operation of ten fundamental principles of being. This conspiration centralized in his visible form to the destruction of the atoms of his body,--not their destruction as substance, but their destruction as material atoms and their reduction to "energy."

The utilization of the principle of vibration will be through the application of electro-magnetic potency concentrated upon a series of octaves of vibrating reeds and the conduction of this through magnets to induced currents. There will be no limit to the lifting power, for weight will be actually destroyed with the proper application of this force to the neutrali-zation of the force of gravity.

SYNOPTICAL OUTLINE OF THE UNIVERSAL EGG OR SHELL

KORESH

THE UNIVERSE is an egg or shell, obtaining as a structure perpetually recreative and existent.

2. It is limited by the environment of its shell, circumferentially, and by the astral nucleus, centrally.

3. Its circumference has seven metallic layers of superimposed strata, five mineral strata, and, above or interior to these, the strata of earth and water.

4. The inner surface of the shell is land and water, comprising a concave surface inhabited by every form and quality of life.

5. Within the environment of the complex circumference of the cell or egg are three distinct atmospheres. Within this is the solar sphere, and within the whole, and constituting its nucleus, is the astral or stellar center.

6. Within the shell and distributed through the three atmospheres are three distinct domains of stars, focalized through the reflecting and refracting powers of the atmospheres and spheres of substance which also, at intervals, occupy the interspace between the center and the circumference.

7. There are seven aggregations of substance, denominated planets, reflected from the strata or laminæ which comprise the outer crust of the environing shell.

8. The stars are not worlds, but focal points of substance or centers of combustion.

9. The planets are not inhabited worlds, but spheres of substance aggregated through the impact of afferent and efferent fluxions of essence proceeding from the circumference and nucleus of the cell.

10. The nucleus is near the center, which is half the diameter of eight thousand miles, the inner surface of the sphere (land and water) constituting the circumference of said diameter. At this astral or stellar nucleus the aggregation resultant from the flow of efferent and afferent substances assumes, naturally, certain geometric lines, curves, and angles, taking the shape of the cone, cube, and oblong in a correlate union of the tent, tabernacle, and courts. This is the description of the astral nucleus of the physical universe, and the geometrical tablature of the materially expressed and unfolded type of the human brain when perfected in the, image and likeness of God.

11. The astral nucleus, owing to the influence of the impinging substances which aggregate and conspire to comprise and determine its form, also assumes motions to correspond with the conspiration and determination of substances.

Motion and Form Are Correlates

1. Motion is not a thing or quality independent of substance.

2. There can be no motion without substance in motion.

3. All motion is the product of combustion of matter.

4. The combustion of matter, through the motion and consequent friction (agitation) of the same, generates spirit.

5. Physical or mental spirit is substance, and is the product of .atomic dissolution.

6. Spirit, though substance, is not matter. Substance is a term applying equally to matter and spirit, whether physical or mental. Spirit and matter are terms to distinguish the two qualities of substance.

7. "Potential energy," so called, though a modern "scientific" term, is not strictly a correct use of language, because where spirit ceases to be kinetic (in motion) it has been deposited as matter.

8. Matter is potential, and when destroyed as matter-substance it becomes spirit-substance.

9. There are two general kinds of alchemico-organic motion: the first is the motion of spirit-substance as spirit-substance; the second, the impulse of matter as an atom, molecule, or mass. The first may be represented by the current of electricity through the wire; the second, by the motion of a ball thrown from the hand, or a projectile from a cannon.

10. Motion of both kinds can be direct, revolutionary, gyrate, and corruscate.

11. Physical or mental spirit is not a mere mode of motion, but substance in motion; and kinds, combinations, and determinations of motion govern the qualities and limitations of form proceeding from the operations of motion.

12. The most complex mind is the correlate of the most complex organic structure. They are so reciprocally related that one cannot, never did, and never will, exist without the other.

13. There can be no greater absurdity than that conception which supposes that spirit, physical or mental, independently of form, could produce anything.

14. Form and function, as matrix and vivifier, exhibited to our inspection in the form and motions of the alchemico-organic universe, never had a beginning, nor will they ever have an ending. The complex world or universe of function, through its varied channels of vitalization, flows into, vitalizes, and is transposed to, the forms and qualities of matter adapted to the genius of its quality, and through the dematerialization of atom, molecule, and complex life, receives again the products of its genius.

Alchemico-Organic World Is a Macrocosm; Its Astral Nucleus Is the Alchemic Microcosm

1. There are two macrocosms; namely, the alchemico-organic and the anthropostic. By this we mean that man, as a universal organism, is a macrocosm in the form and function, as anthropostic, of the alchemico-organic as natural alchemic. In other words, man as a whole is as a shell, rind or pediment of a stellar or astral nucleus, with intermediate organic masses. The Astral Center or nucleus appears epochally, as in Jesus the Christ in the beginning of the Christian era.

2. As the astral nucleus of the alchemico-organic cosmos is the point of congeries of all influent substances, and constitutes the point, contact or touch of all things, looking out to circumferences, and receiving from the same the response of that radiation, so the Astral Center of the anthropostic cosmos, the Man Jesus Christ, or any final manifestation of the Jehovah, is the nucleus or focal point of touch or contact of the human environments of the universal man.

IMPORTANT ADMISSIONS

"When we consider that the advocates of the earth's stationary position can account for and explain the celestial phenomena as accurately, to their own thinking, as we can to ours, in addition to which they have the evidences of their senses, which we have not, and Scriptures and facts in their favor, which we have not, it is not without some show of reason that they maintain the superiority of their system. Whereas, we must be content, at present, to take for granted the truth of the hypothesis of the earth's motion for one thing. We shall never, indeed, arrive at a time when we shall be able to pronounce it absolutely proved to be true. The nature, of the subject excludes such a possibility. However perfect our theory may appear, in our estimation, and however satisfactorily the Newtonian hypothesis may seem to account for all celestial phenomena, yet we are here compelled to admit the astounding truth, that if our premise be disputed and our facts challenged, the whole range of astronomy does not contain the proofs of its own accuracy. Startling as this announcement may appear, it is nevertheless true; and astronomy would indeed be helpless, were it not for the implied approval of those whose authority is considered a guarantee of its truth. Should this sole refuge fail us, all our arguments, all our observations, all our boasted accuracy would be useless, and the whole science of modern astronomy must fall to the ground!"--Dr. Woodhouse, Astronomer, Cambridge, Eng.

Let no one, "so far as hypotheses are concerned, expect anything certain from astronomy, since that science can afford nothing of the kind. The hypothesis of the terrestrial motion of the earth was nothing but an hypothesis, valuable only so far as it explained phenomena, and not to be considered with reference to absolute truth."--Copernicus, Founder of Modern Astronomy.

"In whatever way or manner may have occurred this business, I must still say that I curse this modern theory of cosmogony (the Copernican system), and hope that perchance there may appear in due time some scientist of genius who will pick up courage to upset this universally disseminated delirium of lunatics."--Von Goethe.

"For the invisible things of Him from the creation of the world are clearly seen, being understood by the things that are made, even his eternal power and Godhead." (Rom. i: 20.)

"Who hath measured the waters in the hollow [concavity] of his hand, and meted out heaven with the span, and comprehended the dust of the earth in a measure, weighed the mountains in scales, and the hills in a balance?" (Isa. xl: 12.)

"Where wast thou when I laid the foundations of the earth? . . . Who hast laid the measures thereof? or who hath stretched the line upon it?" (Job xxxviii: 4, 5.)

The "man with a measuring line in his hand," who "stood upon a wall made by a plumbline, with a plumbline in his hand." "He stood, and measured the earth." (Zech. ii: 1; Amos vii: 7, Hab. iii: 6.)

PRINCIPLES OF THE NEW GEODESY

LUCIE PAGE BORDEN

THE ORDINARY READER is not so familiar with the term geodesy as the surveyor, because the latter is in the habit of making use of geodetic instruments in the pursuit of his own vocation. These instruments are designed to aid the mechanic and the civil engineer in laying out public works and surveying large tracts of land, where it is necessary to introduce the question of the earth's curvature.

It is quite enough for the ordinary reader if he be able to define the word geodesy, from its root meaning. Not so with the student of Koreshanity. He knows that the fundamental proposition of this cult calls for a more extended knowledge of the subject than is held by the casual reader. He is therefore required at the outset to acquaint himself with the application of the term. He knows that while he is studying Koreshan Science, it is one of the first requisites to understand what is meant by a geodetic survey.

The most interesting account of the authenticity of the claims made by Koreshan Science would be unintelligible to one who was not aware of the significance of the terms used. The New Geodesy has not departed from any of the known facts of mathematics. It simply rests upon the assertion which nobody can deny, that the chord of an arc is not a tangent. It is certain to all who know them, that the Koreshan geodesists were sincere and indefatigable in their efforts to ascertain whether the direction of the earth's curvature is correctly stated by the ordinary works on surveying.

It was necessary to invent an apparatus which should be capable of determining whether a line extended horizontal to a plumb vertical would prove to be a tangent or a chord. The ordinary surveyor takes it for granted in all his measurements that the surface upon which they are drawn is convex. The Koreshan does not incline to work from an assumption. His method is to use the new apparatus, which is perfectly simple, to settle a question which the ordinary surveyor takes for granted.

It becomes necessary in every extended geodetic operation to devise some method of constructing an instrument to extend a right line. The ordinary surveyor uses the theodolite or engineer's level, and does not understand why there must be an allowance made for perspective foreshortening. He is actually obliged to allow two or three inches to the mile for this factor, but he puts it down as an example of refraction. The Koreshan does not trust to any optical instrument, but tries to make his line absolutely straight by the use of the mechanical principle of right angles.

The civil engineer knows that when he has leveled his telescope or transit instrument, and trains it on his pole a mile away, he will find that the altitude of the cross hair is five or six inches below the height of the pole. Believing that the earth's surface is convex, he is puzzled to know why the difference should not be eight inches to agree with the calculated estimate.

He considers the discrepancy is due to refraction, and claims that he is obliged to allow two or three inches to the mile for this optical factor. The Koreshan geodesist understands the law of perspective foreshortening, and differs from the engineer in his theory of refraction; but the Koreshan prefers to eliminate optical factors and to run a mechanical line.

So much for the differences between the old geodesy and the new. The most intricate problems of surveying are hot unnoticed by the latter, but the out-look of the New Geodesy is so wholly dissimilar to that of the old, that it will be obvious why theoretical problems raised by competent surveyors would be of no interest to the student who has proved to his own satisfaction that the basis of their work is an erroneous assumption.

COPERNICAN HYPOTHESIS IN THE SCHOOLS

LUCIE PAGE BORDEN

THE NOBILITY AND GRANDEUR of the Koreshan premise as it confronts the world are apparent to those who are able to discern the need of the hour. The discoveries of modern investigators in the field of research have only plunged the whole body of so called "scientists" into deeper darkness. The field of astronomical research has been ploughed, but no results have been found to compare with the extension of an air-line to test the contour of the earth's surface and determine in an unmistakable manner that the famous Copernican hypothesis has been superseded by a fact that is known.

What has been taught by the Koreshan System during the past thirty-five years comprises actual knowledge. It does not deal with conjectures nor find any pleasure in drawing base lines for triangulation, when the assumption is that the earth revolves in an orbit about the sun. The conspicuous absence of facts in the old system shows how much time is wasted in searching the heavens, when the same amount of time put into the investigation of the earth's shape would show that it had never been satisfactorily tested until the Koreshan concept was put forth.

To understand the universe is the acme of human desire; for all questions of conduct would be solved by a comprehension of how much is involved in life. The discovery of the nineteenth century was made in 1870, when the actual proof of the Cellular Cosmogony was revealed by the intellect that had become fitted to involve the secrets of creation. Since that date there has been nothing of much importance to herald. Of course the application of forces or energies, frankly admitted to be an impenetrable mystery in their essence, to the uses of life, does go forward without furnishing any conclusive proofs to guide the world to clearer conceptions of being. The electric light has not shown the nature of the electric fluid.

The whole question of putting modern physics into the hands of the young people in our schools and colleges, does not rest upon an assumption according to the ideas of modern educators. But when the subject of electricity is treated, it develops that nothing is known of it, per se. The most that a modern instructor can do is to explain the latest theory, prefacing his remarks perchance with the statement that nothing is absolutely known of the matter under discussion.

Chemistry has held a very important place in the college curriculum. Laboratories have been generously provided to facilitate the study of what was called a science. It has had to revise all its conclusions and to renounce its groundwork in the sight of the indestructible atom removed; and yet, the men who study this branch are not yet willing in more than a few cases to admit the truth of the "discarded science" of Alchemy. They are still trying to find out what has been authoritatively stated by Koreshanity ever since its introduction into the universe of life.

The same fundamental errors which distinguish chemistry, prevent biology from reaching its object. The origin of life, either primarily or proximately, is under discussion daily, without training the mind of the student to apprehend the facts of existence, because he is not taught to know God, in whom is life. The origin and destiny of man are completely misunderstood, and so long as these points are not clear, neither ethics nor the various branches which relate to man can make any progress.

Geology is pivoted on the idea, thoroughly inculcated by the exponents of the Copernican hypothesis; that the earth had a beginning. How far back that epoch should be placed is a moot point; but the educated men of this period in the world's history think it was once bereft of the presence of man, because no human remains are found in certain strata.

Sociology as a science taught in the best colleges is really non-existent. The pattern for construction is wanting, because the solar system is not understood. The principle of organic unity shown in the. cell with its center and environing walls, must be put forward to insure the orderly activity of society. With the fundamental error in social construction expunged, the students might be able to grapple with the labor question and the various branches of economics to some purpose.

Psychology is trying to ferret out the mysteries of modern spiritualism, and is concerned with the measurement of motory and sensory impulses. These are all necessary in their places, but the subject of the soul and the rapport between spirit, soul, and body, are too vast in their bearing to be put aside; yet the nature of the soul is still unknown.

Metaphysics has no conception of the beautiful truths that are brought to light in the study of the brain under the instruction of Koreshanity, whose Founder has shown the intricate harmonies of cell and fibre, as used in the correlated organs of the mind.

The above are some of the reasons why modern education is defective in its instructions. The theorizing of the past has been put up in labeled packages to confuse the mind of the student. If nothing were taught without a groundwork of facts, the mind would not be furnished with old rubbish like an attic or lumber room. Education is the greatest possible attainment of life. It should never be disparaged; but that text-books should exist built upon theories and conjectures is a shame in a progressive age.